Soil Quality
&
Pesticide Residues Analysis

NIPA® GENX ELECTRONIC RESOURCES & SOLUTIONS P. LTD.
New Delhi-110 034

Soil Quality
&
Pesticide Residues Analysis

Mahendra Prasad
Scientist (Soil Science)
ICAR- Indian Grassland and Fodder Research Institute
Jhansi, Uttar Pradesh

Pushpendra Koli
Scientist (Agricultural Chemicals)
ICAR-Indian Grassland and Fodder Research Institute
Jhansi, Uttar Pradesh

NIPA® GENX ELECTRONIC RESOURCES & SOLUTIONS P. LTD.
New Delhi-110 034

NIPA® GENX ELECTRONIC
RESOURCES & SOLUTIONS P. LTD.

101,103, Vikas Surya Plaza, CU Block
L.S.C. Market, Pitam Pura, New Delhi-110 034
Ph : +91 11 27341616, 27341717, 27341718
E-mail: newindiapublishingagency@gmail.com
web: www.nipabooks.com

For customer assistance, please contact
Phone: + 91-11-27 34 17 17
Fax: + 91-11- 27 34 16 16
E-Mail: feedbacks@nipabooks.com

ISBN: 978-81-19254-85-9

Composed and Designed by NIPA®.

Preface

In the recent past soil quality has been considered as one of the most concern for food and fodder security as well as human and livestock health. The goal of maintaining soil and feed quality will not be achieve without precise analysis of soil parameters and pesticide residue. The presence of pesticide residues in animal feed and its animal products is in the focus at the present scenario in the view of WTO. The residues not only affect the public health but also cause economic losses to the livestock industry. Not only these affect the health of livestock and human beings but also affect the quality of animal products. Pesticide residues accumulate in the animals either by direct contact with pesticide or by indirect contact with environment. Pesticides are used in forage crops for disease and pest control and they leave the residues in the feed and fodder consumed by animals.

Common soil analysis methods for monitoring and determining soil quality are fundamental for successful soil management programme. There are well established standard methods for soil and pesticide residue analysis. The results on soil analysis are interpreted with the background of upgrade literatures and reported for valid recommendations. These analyses may help farmers to supplement nutrient to the soil system.

This book will provide a virtual link among agricultural scientists engaged in teaching, research and extension activities. This book depicting analytical methods is a compilation for routine use of research scientists, technician's and students working in the soil and pesticide residue laboratory. Because most soil laboratories now deals with both agricultural and environmental concerns.

Authors

Contents

Section II: Pesticide Residue Analysis

Section I: Soil Quality Analysis

1

Importance of Soil Testing Soil Sampling and Processing

Soil testing involves an estimation of its available nutrient status i.e. the portion or amount of nutrient directly available in soil for subsequent uptake by crop plant. This exercise is commonly referred to as soil testing and is used to arrive at optimum fertilizer application ratio. The need for estimation of available nutrient arises because only a small fraction of what the soil contains is the total nutrient content of the soil. Soil test are calibrated by correlating them with crop response and the result from the basis for making fertilizer recommendations.

Estimation of nutrient contents and forms in materials that are involved in nutrient supply and dynamics is a conical step towards planning scientific nutrient management. In this content, both soil and plant testing information comes out of the interpretation of analysis assumes a greater value when their concentrations and amounts can relate to soil fertility, nutrient availability, plant growth, yield and quality of the crop produce.

Why soil testing?

- Soil testing involves an estimation of its available nutrient status
- It gives the amount of nutrient directly available in soil for subsequent uptake by crop plant.
- Guides to arrive at optimum fertilizer application.
- It is a method of evaluating nutrient status (physico-chemical properties) of the soil i.e. the assessment of the fertility of the soil to determine nutrient deficiencies.
- It is also concerned with environmental quality for the community hazards.

Objectives of soil testing

(i) To evaluate soil fertility and its productivity by the estimation of level of nutrient (Low, Medium, High).

(ii) Grouping of soil for their classification

(iii) To determine the specific soil problem such as an acidity, alkalinity and sodicity if exist. Subsequently giving recommendation for their correction (Lime/Gypsum requirement *etc.*)

(iv) To predict the probability of getting maximum response of crops to fertilizers.

Procedure for soil testing

The procedure for testing the soil to meet these objectives is divided into the following major phases:

- Collection of soil samples and its preparation
- Extraction and determination of nutrients and physico-chemical properties of the soil.
- Interpretation of analytical results.
- Recommendation and follow up of results and evaluation of recommendations.

Advantages of soil testing

1. More rapid method as compare to biological or deficiency symptoms/ plant analysis.
2. One may determine the need of the soil before the planting of crop.
3. To determine the suitability of the soil for laying gardens.
4. Lime problems.
5. Soil survey.

Collection of representative soil sample

Since soil is a very heterogeneous mass and the greatest source of error is usually soil samples itself hence, the soil sample collected should be representative of the area sampled and should also be uniform. Variations in slope, texture, colour, crops grown and management levels must be taken into account. Non-representative locations such as recently fertilized plot, bunds and channel, spot near tree and compost pits etc. must be avoided while sampling. The sample should be taken in zig-zag manner. A representative composite soil samples can be composed of 8 to 20 sub samples from a uniform field. A common error in soil sampling occurs when the top few cm of soil are dry and are not included in the normal sample. Hence, the value of soil test depends on how well the samples represent a field.

Apparatus and materials

Khurpi, spade, augers, plastic bowl, scale, rack, wooden roller mortar and pestle, sieve, polythene,paper, cloth bags, labels, cardboard cartons,aluminium boxes etc.

Soil sampling procedure

1. For sampling of soft and moist soil, the tube auger, spades or khurpi is quite satisfactory for sampling. A screw type auger is more convenient on hard/dry soil.
2. If spade or khurpi is used a V shaped cut mat first be made up to the plough layer. A uniform 15 cm thick slice taken out.
3. First, divide the field into separate units depending on variation in slope, colour, texture, crop growth and management.
4. Remove the debris, rocks, gravels etc. from the surface before collecting soil sample.
5. Make a V shape cut into the soil to a depth of sampling and obtain 2 to 3 cm thick vertical slices along the depth.
6. Collect 10-15 samples randomly in zig-zag manner from each field.
7. The mixed soil sample is coned. The top of the cone is flattened. The cone is divided in four equal parts placing a "+" sign on the top. Two diagonally opposite quarters are taken and thoroughly mixed. This process is repeated until 250-500 g of soil sample is retained.
8. The sample must be kept in a clear cloth or polythene bag.
9. Label it with suitable description and identification marks.
10. Send the soil samples to the soil testing laboratory along with the basic information.

Basic information sheet: The soil sample thus collected must be furnished important information like:

1. Sample number
2. Name and address of the farmers.
3. Details of the field and site. Local name of field, Khasra no etc.
4. Date of sampling
5. Name of crop and variety to be sown

6. Source of irrigation
7. Whether the crop in the subsequent season will be irrigated or un-irrigated.
8. Name of crops and fertilizer used in previous years.
9. Date of harvest of the previous crop.
10. Any other problem observed in the field.

Depth of Sampling

The penetration of plant roots is an important consideration in deciding the depth of sampling. Therefore, the following guidelines may be kept in mind.

S. No.	Crop	Soil sampling depth	
		Inches	cm
1.	Grasses and grasslands	2	5
2.	Rice, finger millet, groundnut, pearl millet, small millets *etc.*(shallow rooted crops)	6	15
3.	Cotton, sugarcane, banana, tapioca, vegetables *etc.* (deep rooted crops)	9	22
4.	Perennial crops, plantations and orchard crops	Three samples at 12, 24 and 36 inches	Three samples at 30, 60 and 90 cm

Preparation of soil sample for testing

1. Spread sample for drying on clean cloth, plastic or brown paper sheet.
2. Remove the stone pieces, roots, leaves & other un-decomposed organic residues from the samples.
3. Large lumps of moist soils should be broken.
4. After air drying the samples should be crushed gently and sieved through a 2 mm sieve.
5. About 250 g of sieved sample should be kept in properly labeled sample bag for testing.

Appropriate time for soil sampling

An ideal time for soil sampling is just after harvest of the rabi fodder crops.

Following point should be kept in mind during collection of soil sampling

1. Remove all debris from surface before collection of soil sample.
2. Take separate sample from the areas of different appearances.
3. In row crop take sample in between rows
5. Keep the sample in a clean bag.
6. Avoid contact of the sample with chemicals, fertilizers or manures.
7. A sample should not be taken from large area (more than 1-2 ha).
8. Sample for micronutrient analysis must be collected by steel or rust free *khurpi*/auger and kept in clean polythene bag.
9. Use clean bags for sample collection. Do not use bags which had earlier contained fertilizer, manure or plant protection chemicals *etc*.
10. Use glass, porcelain jar for long duration storage.

References

Physical and chemical methods of soil and water analysis FAO soils bulletin 10. Food and Agricultural Organisation of United Nations.

Soil Fertility and Fertilizers: Samuel Tisdale and Warner Nelson, Mac millan Pub. New York. Analytical Agricultural Chemistry: S.L. Chopra and J.S. Kanwar

2

Determination of Soil pH and Electrical Conductivity

Determination of pH is actually measurement of hydrogen ions activity in soil-water system. It is defined as negative logarithm of the hydrogen ion activity. Mathematically, it is expressed

as: $pH = - \log a H^+$

The pH value of a soil is an indication of soil reaction *i.e.* acidic, neutral or alkaline. The nutrient availability is governed by soil reaction. It is maximum at neutral pH and decreases with increase in acidity or alkalinity. Thus, pH value gives an idea about the availability of nutrients to plants.

Principle

The pH is a measure of hydrogen or hydroxyl ion activity of the soil-water system. It indicates whether the soil is acidic, neutral or alkaline in reaction. Since crop growth suffers much under both very low (strongly acidic) as well as very high pH (alkaline) conditions.

Now a days, most of the pH meters have single combined electrode. Before measuring the pH of the soil, the instrument has to be calibrated with standard buffer solution of known pH. Since, the pH is also affected by the temperature, hence, the pH meter should be adjusted to the temperature of the solution by temperature correction knob.

Apparatus required

pH meter, balance, beakers, measuring cylinder, spatula, glass rod and washing bottle etc.

Reagents

Standard buffer solutions: These may be of pH 4.0, 7.0 or 9.2 and are prepared by dissolving one standard buffer tablet in 100 ml distilled water, it is necessary to prepare fresh buffer solution after few days. In absence of buffer tablet, a 0.05 M potassium hydrogen phthalate solution can be used which gives a pH of 4.0 (Dissolvc 10.21 g. of A.R. grade potassium hydrogen phthalate in distilled water and dilute to 1 liter. Add 1 ml of chloroform or a crystal of thymol per litre as a preservative).

Procedure

(a) Soil to water ratio of 1:2.5 ($pH_{2.5}$)

- Take 20 g soil in 100 ml beaker and add 50 ml. of distilled water to it.
- The suspension is stirred at a regular interval for 30 minutes.
- Determine the pH by immersing electrodes in suspension.
- For soils containing high salts, the pH should be determined by using 0.01M calcium chloride solution. (Dissolve 0.110 g of $CaCl_2$ in water and dilute to 1 litre).

(b) Saturates soil paste (pHs)

- Add small amount of distilled water to 250g of air dried soil.
- Stir the mixture with a spatula.
- At saturation, the soil paste glistens and flows slightly when the container is tapped it slides freely and ensures cleanly off the spatula.
- After mixing, allow the sample to stand for an hour.
- If the paste has stiffened markedly or lost its glistening, add more water or if free water has collected on the surface of the paste.
- Add an additional weighed quantity of dry soil and mix it again.
- Then insert the electrode carefully in the paste and measure the pH.

(c) Saturation extracts (pHe)

The soil is extracted using vacuum extractor and the pH is measured in the saturation extract.

Categories of soil pH values:

Soil pH		Interpretation
< 5.0	:	Strongly Acidic
5.1– 6-5	:	Slightly Acidic
6.6– 7.5	:	Neutral

7.6– 8.0	:	Mild Alkaline
> 8.0	:	Strongly Alkaline

Electrical Conductivity

Electrical Conductivity is a measure of the ability of a substance/solution to conduct an **Electric Current** (this electric current is carried by ions and the chemical changes that occur in the solution). Amount of soluble salts in a sample is expressed in terms of the electrical conductivity (EC) and measured by a conductivity meter. The Instrument is also available as an already calibrated assembly (Solubridge) for representing the conductivity of solutions in dSm^{-1} (deci Siemen per meter) at 25⁰C.

Principle

Salt ions, like metals, allow the electric current to pass through them. Hence, the electrical conductivity (EC) of the soil water system rises with increasing content of the soluble salts in the soil.

Apparatus required

Conductivity meter, balance, beakers, measuring cylinder, spatula, glass rod and washing bottle etc.

Reagents

Potassium chloride: Dissolve 0.7456g dry potassium chloride (AR) in distilled water and make up the volume to one litre.

Procedure

- Take 20 g of soil in 100 ml beaker.
- Add 50 ml of distilled water and shake intermittently for 30 minutes.
- Determine the conductivity of the supernatant liquid with the help of conductivity meter.
- The electrical conductivity of saturation extract (E.C.e) is also determined for salinity ratings.

EC($dS\ m^{-1}$)	Effect
<1 -	No deleterious effect on crop
1-2 -	Critical for salt sensitive crops
2-3 -	Critical for salt tolerant crops
>3 -	Injurious to most crops

3

Determination of Soil Texture by Hydrometer Method

The method is based on the Stokes law which states that the rate of fall of particles in suspension is directly proportional to their size. Accordingly, larger particle (sand) will settle first, followed by silt and clay. The resistance offered by the liquid to the fall of the particles varies with the radius of the sphere.

Principle

This method is based on the principle of dispersion and sedimentation techniques, employed to a given weight of soil sample. Mechanical stirring disperses macro aggregates while chemical dispersing required for micro aggregates. The mixture of dispersed soil particles in water is called a soil suspension. Sedimentation refers to the settling rates of the dispersed particles in water, which is a function of particle size and is governed by Stoke's law.

Equipment and Apparatus Required

a. Hydrometer

b. Electrically driven mixer

c. Graduated cylinder (1L)

d. Beaker

e. Hot plate

f. Watch glass

g. Thermometer

Reagents

a. 5% sodium hexametaphoshphate

b. Hydrogen peroxide

Removing of cementing agents

(a) **Organic matter**: it is removed with the hydrogen peroxide treatment of soil. Hydrogen peroxide oxidizes the organic matter present as cementing agent to carbon dioxide and water.

(b) **Dehydration and physical separation**: clay themselves act as cementing agents. Dehydration of clays brings particles together, due to strong cohesive forces, developing between particles.

(c) **Oxides of iron and aluminum**: the removal of iron and aluminium is very important in lateritic soils. The oxides are reduced to soluble forms at the cost of acceptance of electrons from the reducing agents viz. sodium dithionate in sodium acetate.

(d) **Calcium carbonate**: Cementation due to calcium carbonate is removed by treating the soil with dilute HCl.

Procedure

- Pass the soil through a 2 mm sieve. Weigh 50 g fine textured soil or 100 g coarse textured soil.
- Add 50 to 60 ml of 6% hydrogen peroxide, cover the beaker with watch glass and placed it on a water bath until the organic matter id oxidized. Remove the beaker and allow it to cool.
- Repeat the process until frothing stops.
- Transfer the contents into a dispersing cup, with about 400 ml of distilled water.
- Add 100 ml of sodium hexametaphosphate solution into it.
- Stir the suspension with the help of an electric stirrer for at least 10 minutes.
- Transfer the suspension into a settling cylinder and make the suspension upto the 1 litre mark.
- Place a rubber stopper over the mouth of the cylinder and shake it vigorously back and forth for 1 min.
- Place the beaker on a table and note the time immediately.
- Insert the hydrometer into the suspension and take the first reading after 4 minutes when the particles >0.02 mm have settled. If the surface of the suspension is frothy, add 1 drop of amyl alcohol.

- Remove the hydrometer carefully and wash it with distilled water. Record the temperature of the suspension.
- Allow the suspension to remain undisturbed and reinsert the hydrometer at the end of 2 hours, after the initial shaking was stopped. Now the particles >0.002 mm mm i.e. sand +silt have settled. Record the hydrometer reading.
- Calculate the sand, silt and clay composition and determine the texture using ISSS textural triangle.

How to calculate corrected hydrometer reading

i.e. Let a 4 minute reading of 15 at 77 °F be obtained when a 50 g oven dry soil sample is used.

Corrected hydrometer reading= 15+(77-67)×0.2=17

Percent (silt+clay) in the suspension=(17/50)=34

(The hydrometer is calibrated at 67 °F, if the working tem is above 67 °F, the correction is added, and if below, the correction factor is subtracted.)

Observations and calculation

a. Weight of soil (g)=

b. Hydrometer reading at 4 min (g) =

c. Tem of the suspension at 4 min observation=

d. Corrected hydrometer reading at 4 min(g)=

e. Hydrometer reading at 2 hrs (g) =

f. Tem of the suspension at 2 hrs observation=

g. Corrected hydrometer reading at 2 hrs=

h. Amount of silt+clay(g)=d

i. Amount of clay (g)=g

j. Amount of silt(g)=d-g

k. % silt=d-g/a×100

l. % clay=g/a×100

m. %sand=100-(k+l)

The texture of the soil is determined from the relative distribution of sand, silt and clay in the sample. Triangular classification (Fig. 1), involving 12 different textural classes are use for the determination of the textural class of the test sample.

For using the triangular classification, take the following steps:

1. Locate the percent clay content of the sample on the left hand side and move horizontally.
2. Locate the percent silt content of the sample on the right hand side and move on slanting line. Alternatively, sand content instead of silt may be located.
3. Find the point where two lines cross each other to get the textural class.

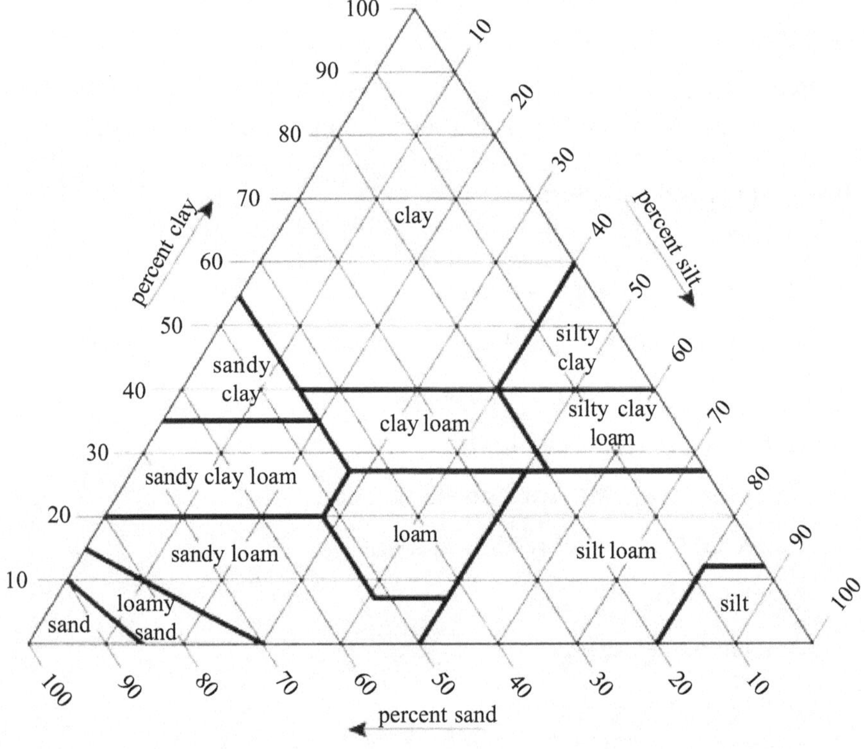

Fig. 1: Textural triangular showing the content of sand, silt and clay (%)

4

Determination of Soil Bulk Density

Purpose

The bulk density varies indirectly with total pore space present in the soil and gives a good estimate of porosity of soil. Bulk density is of great importance than particle density in understanding the physical condition of soil.

Soil bulk density is defined as the ratio of the mass of the oven dry soil to its bulk volume.

Method of determination

Different methods are available for bulk density determination which differ how the soil sample is obtained and its volume is determined such as core, clod and excavation methods. There are other methods for the determination of bulk density where different principles are employed, for example, radiation method. Commonly used methods are weighing bottle, core and clod method.

Core Method

The method involves sampling a core from a desired depth, in its most natural condition using core sampler and determining the mass of the solids and the water content of the core by weighing the wet core, drying it to constant weight in an oven at 105°C and weight. Bulk density is then calculated from measurement of the bulk volume using the core length and the diameter of the cutting edge of the sampler.

Apparatus required

Core sampler with removable sample cylinder, shovel, straight edge knife, moisture box, cellophane tap, balance and oven

Procedure

- Select two clean and crack free smooth soil surface for duplicate sampling.
- Drive the core sampler vertically into level ground, deep enough, to fit the sampler can in the sampler.

- Dig out the sampler by means of a *khurpi* or spade, and remove the sample can without disturbing the natural structure of soil.
- Trim off the extra soil from both ends of the sample can with a sharp knife.
- Transfer the soil sample in a moisture box and seal with cellophane tap for transportation to the laboratory.
- In the laboratory, remove the tape, weigh the sample, uncap and place in oven at 105°C for drying until constant weight.
- Calculate bulk density along with moisture content in the sample.

Observations and Calculations

A. Date

B. Field number or name

C. Volume of soil sample = 3.14 ×(radius of sample can)^2× height of sample can

$$\text{Bulk density}\left(\text{g cm}^{-3}\text{or Mg m}^{-3}\right)=\frac{\text{Oven dry weight of soil sample (g)}}{\text{Volume of soil sample}\left(\text{m}^3\right)}$$

$$\text{Moisture content in soil}\ (\%)=\frac{\text{Fresh weight of soil}-\text{Oven dry weight of soil}}{\text{Oven dry weight of soil}}\times 100$$

Weighing bottle method (Pycnometer method)

Principle

The mass of the soil is determined by weighing the oven dry soil sample. The soil in small amounts, say 10 gm., is placed in a container which is tapped 15-20 times on a table be letting it fall from a height of about 2-3 cm. This tapping is assumed to produce the same packing as occurring naturally in the field, even though this assumption is not strictly correct. The volume of this packed soil will be equal to the volume of the container. Bulk density is calculated from the mass and volume of the soil.

Apparatus required

a. A weighing bottle (50 ml)

b. Balance and

c. Aburette of 50 ml capacity

Procedure

- Weigh an empty 50 ml bottle.
- Fill the bottle with oven dry soil upto the brim by tapping and weigh it.
- Empty the bottle and determine its exact volume using burette.

Observations and calculations

a. Mass of empty bottle = M1 gm.

b. Mass of bottle + soil = M2 gm.

c. Mass of the soil = (M2 – M1) gm.

d. Volume of water filling the bottle = V cm3.

Bulk density = (M2 – M1) / V g cm^{-3} or Mg m^{-3}

5

Determination of Soil Particle Density

Particle density determination is based on the measurements of weight of solids and their volume. The weight of solids is obtained by weighing a sample of oven dried soil. The volume of solids is determined is determined by immersion. The method outlined here is "Pycnometer method".

Principle

A given amount of dry soil when immerses in a definitive volume of water, expels air and results in the displacement of equal volume of water. The volume of soil particles is determined by measuring the volume of water displaced in the pycnometer bottle.

Apparatus required

a. Pycnometer

b. Pipette

c. Balance

d. Water bath or hot plate

e. Filter paper

Procedure

- Fill up a clean, dry pycnometer with water.
- Replace the stopper and wipe out the surface of the pycnometer and weigh it
- Empty it and put into a 10 g soil
- When bottled is half filled, using the pipette, wash into it soil particles sticking to the inner side of the neck.
- Boil the contents to remove the entrapped air.

- Cool the contents to room temperature and fill the pycnometer with water.
- Insert the stopper, wipe the surface of the pycnometer, dry and weigh it.

Observations & Calculations:

a. Weight of water filled pycnometer = A g

b. Wt of dry soil= 10 g

c. Wt. of pycnometer + water + soil =B

d. Volume of water displaced =(A+10-B) cm^3

e. Particle density of soil= Mg cm^{-3}

$$\text{Particle density } (g\ cm^{-3} \text{ or } Mg\ m^{-3}) = \frac{10}{A+10-B}$$

6

Determination of Water Holding Capacity in Soil

Purpose

The determination of water holding capacity in soils is important as it gives an idea of the capacity of soil to hold water for the use by crops. The light soils which do not hold such water require more frequent irrigations than heavy clay soils; well decomposed organic matter increases the water holding capacity. Exchangeable sodium and type of clay mineral also have a marked effect on water holding capacity. Water holding capacity of soils is useful for selection of soils for irrigability classification. It also helps for comparing other properties of soils.

Apparatus required

a. Physical balance

b. Keen's boxes

c. Enamel tray.

d. Petri dish

e. Filter paper

f. Hot air oven

Procedure

- Crush air-dry soil and pass through 2 mm sieve.
- Place round filter paper and fix it to the internal perforated floor of the dish. The weight of the dish and filter paper is noted. The dish is then filled with soil by tapping the dish briskly & making plane the top of soil and find out its weight.

- Place the set of perforated dishes in enamel tray. Pour the water in enamel tray at half of height of dish. Water may rise in dish through perforated bottom and moist the soil to its capacity. Keep it for 5 to 6 hours in water.
- Take the dishes and place it on a filter paper sheet, so that the excess of water may drain away from the pores within half an hour. The dish containing moist soil is weighted and the weight is noted.

Observations

a. Weight of empty dish + filter paper =W1 gms.

b. Weight of empty dish + filter paper=W2 gms.

   + oven dry soil

c. Weight of empty dish + filter paper =W3 gms.

   + wet soil

d. Weight of moisture absorbed by the filter paper=W4

Calculations

WHC(%)=(W3-W2-W4/W2-W1)×100

7

Determination of Water Stable Aggregates of Soil by Wet Sieving Method

Soil aggregates refers to a group of two or more primary particles which cohere to each other more strongly than surrounding particles. Under field condition, all such units also cohere to some degree with neighboring aggregates and form larger aggregates. Such aggregates form and break easily into smaller aggregates during tillage and by disruptive action of water and air. The aggregates maintaining their identity are those in which the cohesive forces among particles are greater than the disruptive forces. A quantitative characterization of aggregates is done by determining aggregate stability and size distribution of aggregates. The aggregate stability refers to the resistance of soil aggregates to breakdown by water, air and mechanical manipulations.

The soil aggregates which can stand the disruptive forces of wetting by water are known as water stable aggregates. Therefore, the process of wetting, disruption of dry aggregates and screening of aggregates of different sizes should reasonably compare to the disruptive action of water and mechanical forces of tillage under wet conditions. The vacuum wetting of dry soil largely simulates the process of soil wetting in the surface layers but the immersion wetting is more comparable to the wetting of surface soil by irrigation. Similarly, sieving under water compares more closely with the disruptive action of water and other mechanical forces. The mean weight diameter (MWD) of water stable aggregates is a useful index to characterize the change in structure due to tillage. The mean weight diameter can be calculated using following equation.

$$MWD = \sum_{i=1}^{n} Xi\ Wi$$

Where Xi= mean diameter of size fraction I, and Wi= proportion of total sample weight occurring in the corresponding size fraction i. The summation is carried out over all n size fractions including the one that passes through the finest sieve.

Principle

The wet sieving method involves a given amount of soil aggregates in a nest of standard sieves in water for a given length of time, followed by the collection of aggregates plus coarse material retained on each sieve and their weight. Finally, the soil mass, retained on each sieve is dispersed in H_2O_2 and HCl and passed through individual sieves for the coarse soil fractions, which otherwise might be included wrongly while reporting the mean weight diameter and the percent of the total aggregates in different size fractions of the soil mass.

Apparatus Required

Standard sieves -2 sets (8.0,5.0,1.0,0.5,0.25 and 0.1 mm), Yodar apparatus (which raises and lowers the nest of sieves through water 3.8 cm, approx. 30 times per minute) electronic balance, oven, watch glass etc.

Reagents

H_2O_2 and 0.1N HCl

Procedure

- Take 200 g of air dry soil which can pass through 8.0 mm sieve and retained on 5.0 mm screen. Do not break them too small. Large gravels or roots should be removed.
- Weight 25 g aggregates retained on 5.0 mm sieve separately in three watch glasses. Keep one of them in oven at 105°C for moisture determination and use the other two for analysis in duplicate.
- Arrange two set of six sieves in the order of 5.0,2.0, 1.0, 0.5,0.25, and 0.1 mm from top to bottom.
- Spread the sample (25g) of aggregates evenly over top 5 mm sieve and spray 5 to 10 ml of salt free water on them. Wait for 2 to 5 minutes.
- Transfer the nest of sieves to the drum of sieve shaker and clamp them in position. Fill the drum, with salt free water up to a level slightly below the top screen when sieves are in highest position. Turn the pulley of the shaker slowly by hand to attain the highest position.
- Lower the sieves to the lowest position and wet the aggregates for 10 minutes. Fill more water in the drum so that aggregates are just covered with water when sieves are again in highest position.
- Switch on the oscillator and let the sieve oscillate in water for 10 minutes with a frequency of 30 cycles per minute through a stroke length of about 3.8 cm.

- Take out the nest of sieve, let the water drain, separate and place them on paper sheets, and let the aggregates on each sieve dry and harden in air.
- Transfer the aggregates of each sieve in to separate moisture boxes. Find out their oven dry weight.
- Transfer the aggregates from moisture boxes into 250 ml beakers separately and oxidize and disperse using H_2O_2 and sodium hexametaphosphate respectively. Use mechanical stirrer for complete dispersion; pass the dispersed aggregates through the same sieves on which they were retained originally. Collect the un-aggregated primary particles from each sieve and record their oven dry weight. Subtract the weight of the primary particles from the weight of the aggregates obtained on respective sieves in previous step.
- Calculate the percentage of aggregated soil particles on different sieves.
- Plot a graph between accumulated percentage of soil remaining on each sieve as ordinate and upper limit of each size fraction as abscissa.
- Calculate the mean weight diameter (MWD) of aggregates in mm and report the results as MWD and percent aggregation.

Observations

a. Air dry weight of sample (g)=

b. Moisture percentage in soil (%)=

c. Frequency of oscillation (min^{-1})=

d. Stroke length (mm)=

e. Oven dry weight of particles i) aggregated(g)=

ii)Un-aggregated(g)=

Oven dry weight of sample (g)=

Result

1. Percentage aggregates greater than 0.1 mm (mean of 2 observations)=
2. MWD(mean of 2 observations from calculation)=
3. MWD(mean of 2 observations from graphs)=

Calculations

Sl. No.	Size range of aggregates (mm)	Weight of particles retained on the sieve (g)				Wt. of aggregated particle		% aggregate		Accumulated percentage	
		Before dispersion		After dispersion		I Sample	II Sample	I Sample	II Sample	I Sample	II Sample
		I Sample	II Sample	I Sample	II Sample	c-d		(e×100)/f			
a	b	c		d		e		F		g	
1.	8.0-5.0										
2.	5.0-2.0										
3.	2.0-1.0										
4.	1.0-0.5										
5.	0.5-0.25										
6.	0.25-0.10										

Mean Weight Diameter $\sum_{i=1}^{n} Xi\,Wi\,mm$ mm

Where n=6, the number of size fractions.

8

Determination of Soil Organic Carbon

Soil Organic Matter (SOM) is the seat of nitrogen in soil and its determination is often carried out as an index of nitrogen availability. The top soil (surface soil) range from 0.5% to 3.0% organic carbon for most upland soils and less than 0.5% are limited to desert soils. Soil organic matter averages about 58% carbon, it follows that soils generally range from about 2 to 6 % organic matter (% O.M. = %C x 1.724, The factor 1.724 = 100/58). There is also a close relationship between carbon and nitrogen in soils. Most organic matter average about 5% nitrogen so that the N: C ratio is 1:11.6. Therefore by multiplying the soil organic matter percentage by 0.05 an approximate value for the soil nitrogen, percentage is obtained. In soil the chief source of some of the nutrients essential for plant growth is organic matter, such nutrients are N, S and boron is also largely derived from organic matter.

Principle

A known weight of soil is treated with the excess volume of standard $K_2Cr_2O_7$ in the presence of conc. H_2SO_4. The soil is slowly digested at the low temperature, by the heat of dilution of $H_2SO_4$and organic carbon in the soil is thus oxidized to CO_2. The excess of $K_2Cr_2O_7$not reduced by the organic matter is titrated back against a standard solution of ferrous ammonium sulphate in the presence of phosphoric acid and diphylamine indicator.

1. Oxidation

   $$2\ K_2Cr_2O_7 + 8\ H_2SO_4 + 3\ C \rightarrow 2K_2SO_4 + 2\ Cr_2(SO_4)_3 + 8\ H_2O + 3CO_2$$

2. Titration

   $$2FeSO_4(NH_4).12H_2O + H_2SO_4 + O \rightarrow 2(NH_4)_2SO_4 + Fe_2(SO_4)_3 + 13\ H_2O$$

Apparatus and Reagents

a. 500 ml conical flasks.

b. Pipette

c. Burette

d. Phosphoric acid 85%.

e. Sulphuric acid 96 %.

f. Standard 1*N* $K_2Cr_2O_7$ – 49.04 g/liter.

g. Standard 0.5 *N* Fe $(NH_4)_2 (SO_4)_2 . 6H_2O$ 196 g in 800 ml water containing 20 ml H_2SO_4 and diluted to1 litre. \

h. Diphenylamine – 0.5g in 20 ml water and add 100 ml Conc. H_2SO_4.

Procedure

- Weigh 1g soil sample in 500 ml conical flask. Add 10 ml of 1 N $K_2Cr_2O_7$ and 20 ml conc. H_2SO_4. Mix thoroughly and allow reaction to proceed for 30 minutes.
- Dilute the reaction mixture with 200 ml water and add 10 H_3PO_4 and 1 ml of diphenylamine.
- Titrate the solution with standard FAS (0.5N) to a brilliant green colour. A blank without soil should be run simultaneously.

Observations and calculation

1. Weight of soil sample- 1 g
2. Normality of $K_2Cr_2O_7$ used-1N
3. Volume of $K_2Cr_2O_7$- 10 ml
4. Normality of FAS- 0.5 N

$$\text{Organic carbon (\%) in soil} = \frac{10(\text{B-S})}{\text{B}} \times 0.003 \times \frac{100}{\text{Wt. of sample}}$$

Organic carbon in soil (%) = OC×1.3

There is incomplete oxidation of organic matter in this procedure. The organic carbon is multiplied by 1.3 on the assumption that there is 77% recovery.

Organic matter in soil = OC(%)×1.724

Assume that organic matter contains 58% carbon. So, 100/58

Interpretation

Organic carbon (%)	Soil rating
<0.5	Low
0.5-0.75	Medium
>0.75	High

9

Determination of Available Nitrogen in Soil

The major part of soil nitrogen exists in complex combination in the organic compounds. It becomes available to plants after breakdown to simple form followed by mineralization. The chemical methods for determination of soil nitrogen availability involve either direct or indirect of total nitrogen or measurement of a fraction of easily hydrosable nitrogen by employing solution of dilute acid or alkali. Distillation of alkaline potassium permanganate solution has often been adopted for estimating the ox disable and reactive form of soil nitrogen. The available nitrogen in soil is determined by alkaline permanganate method as per procedure suggested by Subbiah and Asija (1956).

Principle

A known weight of the soil is mixed with alkaline potassium permanganate ($KMnO_4$) solution and distilled. The organic matter present in soil is oxidized by the nascent oxygen, liberated by potassium permanganate, in the presence of sodium hydroxide and the released ammonia is condensed and absorbed in known volume of a boric acid with mix indicator to form ammonium borate, the excess of which is titrated with a standard sulphuric acid.

Reactions involved

I. Distillation

$2\ KMnO_4 + H_2O \rightarrow 2\ MnO_4 + 2KOH + 3O^-$

Medium(Nascent oxygen)

Oxidation

$R.CHNH_2COOH + O^- \rightarrow R.CO.COOH + NH_3$

Organic- N fraction (Ammonia)

Distillation

$NH_3 + H_2O \rightarrow NH_4OH$

Absorption

$3NH_4OH + H_3BO_3 \rightarrow (NH_4)_3BO_3 + 3H_2O$

(Green colour)

II. Titration

$2(NH_4)_3BO_3 + 3H_2SO_4 \rightarrow 2(NH_4)_3SO_4 + 2\ H_3BO_3$

(Pink colour and original)

Apparatus Required

KEL PLUS automatic nitrogen estimation system, electronic balance, burette conical flask etc.

Reagents

a. 0.32 % potassium permanganate ($KMnO_4$) solution: Dissolve 3.2 g of potassium permanganate in distilled water and make up the volume 1 litre.

b. 2.5 % sodium hydroxide (NaOH): Dissolve 2.5 g of NaOH pellets in distilled water and make up the volume to 1 litre.

c. 2 % boric acid solution containing 20 - 25 ml of mixed indicator / liter.

d. Mixed indicator: 0.066g methyl red + 0.099g bromocresol green dissolve in 100 ml of 95 % alcohol.

e. 0.02 N sulphuric acid (H_2SO_4): Prepare approximately 0.1N H_2SO_4 by adding 2.8 ml of conc. H_2SO_4 to about 1 litre of distilled water. From this prepare 0.02 N H_2SO_4 by diluting a suitable volume five times with distilled water.

Procedure

- Weigh 5 g of prepared soil sample and transfer it to the digestion tube.
- Load the tube in distillation unit and other sides of hose keep 20 ml of 2 % boric acid with mixed indicator in 250 ml conical flask.
- 25 ml each of potassium permanganate (0.32 %) and sodium hydroxide (2.5 %) solution is automatically added by distillation unit programme.

- The sample is heated by passing steam at a steady rate and the liberated ammonia absorbed in 20 ml of 2 % boric acid containing mixed indicator solution kept in a 250 ml conical flask.
- With the absorption of ammonia, the pinkish colour turns to green.
- Nearly 150 ml of distillate is collected in about 10 minutes.
- The green colour distillate is titrating with 0.02N sulphuric acid and the colour changes to original shade (pinkish color).
- Simultaneously, blank sample (without soil) is to be run.
- Note the blank & sample titer reading (ml) and calculate the available nitrogen in soil.

Calculations

$$\text{Available soil N (kg ha}^{-1}) = \frac{\text{(S-B)} \times \text{Normality of acid} \times \text{Atomic wt of N} \times \text{Wt of 1 hactre of soil (0-15 cm)}}{\text{Weight of soil (g)} \times 1000}$$

$$\frac{(S-B) \times 0.02 \times 14 \times 2.24 \times 10^6}{5 \times 1000}$$

Factor= (S-B)×125.44

Where

S= Sample reading (ml)

B=Blank reading (ml)

Interpretation

Available soil N (kg ha^{-1})	Soil rating
<280	Low
280-560	Medium
>560	High

10

Determination of Available Phosphorous in Soil

The term available phosphorus (P) refers to the inorganic form, occurring in soil solution, which is almost exclusively 'orthophosphate'. The available P is considered to be a fairly good indicator/measure of the P supplying capacity of a soil. A knowledge of available P content of soils is important for determining its critical limit , based on soil-test-crop response calibration study.

Principle

In soil, phosphorus exists in the form of various types of orthophosphates. A very small fraction of these is available to plants at a given time. Available phosphorus content of soil consists mainly of Ca-P, Al-P and Fe-P. In the neutral and alkaline soils particularly, Ca-P is the dominant fraction. Organic –P fraction is also considerable amount, but is usually not include in the determination of available phosphorus. A large number of extractions, buffer solutions, acids and chelating agents have been suggested for available phosphorus from time to time. However, no single extractant appears to be suitable for all types of soils. Two types of extraction methods are more popularly adopted. Under acidic soil conditions, Bray No.1, which involves soil extraction with a solution consisting of 0.003 N NH_4F and 0.025 N HCl is widely followed. The fluoride complexes Al and Fe in soil, thus releasing some bound p besides the easily acid soluble P (largely Ca-P). This extraction is suitable for soils containing less than 2% calcite or dolomite because in calcareous soils, carbonates quickly neutralize the acid, resulting in less extraction of P.

The other and most widely used extractant is the 0.5M $NaHCO_3$ solution at pH 8.5. The reagent is most suitable for neutral to alkaline soils and is designated to control the ionic activity of calcium through solubility product of $CaCO_3$, thus extracting the most reactive forms of P from Al-, Fe- and Ca-phosphate. Phosphorus in the extract can be determined using a suitable method of colour development and measuring the colour intensity at an appropriate wavelength.

Bray's No. 1

Instruments

a. Mechanical shaker

b. Spectrophotometer

Reagents

a. Bray's P-1 extractant: Dissolve 1.110 g AR grade ammonium fluoride in one litre of 0.025N HCl.

b. 1.5% Dickman and Bray's reagent: Dissolve 15 g of AR grade ammonium molybdate in300mL of warm water, cool and add exact 350 mL of 10 N HCl. Make the volume to 1 L.

c. 40% $SnCl_2$ stock solution: Weight 10 g pure stannous chloride in a 100 mL glass beaker. Add 25 mL of conc. HCl and dissolve by heating. Cool, transfer to an amber coloured bottle and store in dark after adding a small piece of Zn metal to prevent oxidation. From this, prepare a dilute $SnCl_2$ solution (0.5 mL diluted to 66 mL) immediately before use.

d. Standard stock solution: Weigh 0.439 g AR grade KH_2PO_4 dried in a oven at 60°C for 1 hour in a one litre beaker, add about 500 mL of distilled water and dissolve. Add 25 mL of approx. 7 N H_2SO_4 and make the volume to one litre. This is 100 mg P L^{-1} solution.

e. Standard working solution: Dilute a suitable volume of 100 mg P L^{-1} solution 50 times to get 2 mg P L^{-1} solution.

Procedure

- Weigh 2.5 g of soil sample in a 150 mL conical flask.
- Add 50 mL of Bray P-1 extractant and shake for 5 minutes.
- Filter through whatman No.1 filter paper quickly so as to collect the filtrate within 10 minutes.
- Transfer 5 mL aliquot into a 25 mL volumetric flask.
- Add 5 mL of ammonium molybdate solution shake a little and dilute to about 22 mL.
- Add 1 mL of dilute $SnCl_2$ mix by shaking a little, and make up the volume.
- Run a blank without soil under identical conditions.
- Measure the intensity of the blue colour developed, using 660 nm wavelength (red filter).

Olsen's method with ascorbic acid

Phosphorus is extracted from the soil with 0.5 m $NaHCO_3$ at a nearly constant pH of 8.5. The phosphate ion in solution treated with ascorbic acid in an acidic medium provides a blue colour complex. Measurement of the quantitative determination of phosphorous in soil (Olsen's *et al.*, 1954)

Reagents

a. 0.5 M Sodium bicarbonate ($NaHCO_3$) solution: Dissolve 42g of $NaHCO_3$ in distilled water to get one litre solution and adjust the pH of the solution to 8.5 by small quantity of NaOH.

b. Activated Charcoal: Darco G-60 (P-Free)

c. 5 N Sulphuric acid (H_2SO_4) Solution: Add 141 ml of con. H_2SO_4 to 800 ml of distilled water. Cool the solution and dilute to one litre with distilled water.

d. Reagent A:

   i. Dissolve 12.00 g of ammonium paramolybdate in 250 ml of distilled water.

   ii. Dissolve 0.2908 g of potassium antimony tartrate ($KSbO.C_4H_4O_6$) in 100 ml distilled water.

   iii. Above both solution mix thoroughly and made one litre in volumetric flask with the help of distilled water.

   iv. Add these dissolved reagents to one litre of 5N H_2SO_4.

e. Ascorbic acid working solution (Reagent B):Dissolve 1.056 g of ascorbic acid in 200 ml of reagent A and mix. This ascorbic acid (reagent B) should be prepared as required because it does not keep more than 24 hours.

f. Standard phosphate solution: Weigh 0.4393 g of potassium dihydrogen phosphate (KH_2PO_4) into one litre volumetric flask. Add 500 ml of distilled water and shake the contents until the salt dissolves. Dilute the solution to one litre with distilled water to get 100 ppm P solution. Dilute 20 ml of 100 ppm P solution to one litre to get form-working solution of 2 ppm.

Preparation of standard curve

i. Take different concentration of P (0, 1, 2, 3, 4, 5, etc ml of 2 ppm standard P Solution) in 25 ml volumetric flasks.

ii. Add 5 ml of the 0.5M $NaHCO_3$ extracting solution to each flask, and acidify with 5N H_2SO_4 drop by drop.

iii. Add about 10 ml distilled water and 4 ml of reagent 'B', then shake the solution.

iv. Make the volume 25 ml by distilled water.

v. The intensity of blue colour is read on spectrophotometer at 660 nm wavelengths after 10 minutes.

vi. Plot the curve by taking P concentration on X axis and colorimeter reading on Y axis. Repeat the process till you get straight line relationship.

vii. Calculate the factor i.e. 1 colorimeter reading is equal to how much ppm of phosphorus?

Procedure

- Take 2.5 g of soil sample in 150 ml conical flask and 0.5 g Darco G-60 activated charcoal.
- Then add 50 ml of 0.5 M $NaHCO_3$ solution and shake the solution for 30 minute in a shaker. Similar processes run for a blank without soil.
- Filter the suspension through the Whatman no. 40 paper.
- Take 5 ml aliquot of the extract in a 25 ml volumetric flask, and acidify with 5N H_2SO_4.
- Add small quantity of distilled water, and then add 4 ml of reagent B.
- The intensity of blue colour is read on spectrophotometer at 660 nm wavelengths after 10 minutes.

Observations and Calculations

a. Weight of the soil = 2.5 g

b. Volume of extract=50 ml

c. First dilution=50/2.5=20 times

d. Volume of aliquot taken=5

e. Final volume make up=25

f. Second dilution=25/5= 5times

g. Total dilution=20x5=100 times

h. Conc. from standard curve= R

i. Available P in soil (ppm)=Rx dilution factor

j. Available P in soil (kg/ha)=ppm ×dilution factor×2.24

Interpretation

Available soil P_2O_5(kg ha^{-1})	Soil rating
<22.5	Low
22.5-56	Medium
>56	High

11

Determination of Available Potassium in Soil

The total potassium content (K) of a soil varies from 0.05 to 2.5%. The total K is distributed in mineral form (90-98%), fixed non-exchangeable (1-10%) and exchangeable plus water soluble K(1-2). Exchangeable K represents an average of 2-15% of the sum of exchangeable bases and 1-3% of the total capacity of cationic exchange.

Principle

It is based on the principle of the equilibrium of the soils with an exchangeable cation mode of the solution of neutral ammonium acetate, in a given soil:solution ratio. During the equilibrium, ammonium ions exchange with the exchangeable K ions of soil.The K content in the equilibrium solution is estimated with a flame photometer.

Clay]K^+ CH3COONH_4................. Clay]NH_4^++ CH_3COOK

Apparatus Required

a. 150 ml conical flask

b. Funnel

c. Balance

d. Pipette

e. Filter Paper

f. Mechanical shaker

Reagents

a. 1N CH_3COONH_4 solution: Dissolve 77.09 g of Ammonium acetate in distilled water and make up the volume to 1 litre. Adjust the solution pH to 7.0, adding acetic acid or NH_4OH solution as required.

b. Potassium chloride solution: Dissolve 1.908 g of KCl in distilled water, and make up to the volume to 1 litre. It gives 1000 ppm k solution and is treates as the stock solution of K.

c. Standard curve for K(working standards): From the stock solution, take measured aliquots and dilute with ammonium acetate solution/distilled water to give 10 to 40 ppm of K.

Procedure

- A 150 ml conical flask is taken and washed properly.
- 5 g soil sample is taken in flask.
- 25 ml of 1N $CH3COONH_4$ solution is added to the flask.
- The content in the flask is shaken on a mechanical shaker for 5 minutes.
- Now the content of flask is filtered through filter paper.
- The extract is collected to estimate K.
- Concentration of K is observed by microprocessor based flame photometer by feeding the extract to photometer.

Calculations

- Weight of the soil = 5 g
- Volume of the neutral ammonium Acetate=25 ml
- Dilution factor= 25/5=5
- Available K in soil (ppm)=Rx dilution factor
- Available K in soil (kg/ha)=ppm ×dilution factor×2.24

Interpretation

Available soil K (kg ha^{-1})	Soil rating
<137	Low
137-338	Medium
>338	High

12

Determination of Available Sulphur in Soil

Sulphur (S) occurs in different forms in soil *viz*., sulphites, sulphates , sulphides and in organic compounds. Plants absorb S in the form of sulphate(SO_4^{2-}) . The organic forms of S compounds become assimilable, especially following microbiological transformation into SO_4^{2-}. The mineralization of organic sulphur in a soil depends primarily upon the N:S ratio and any SO_4^{2-} formed may be fixed against extraction, particularly if much Fe and Ba is present or if soil is very acidic.

Principle

Beside some amount in the soil solution, available sulphur in mineral soils occurs mainly as adsorbed SO_4^{2-} ions. Both $CaCl_2$ and phosphate solutions (as monocalcium phosphate) are generally used for replacement of the adsorbed SO_4^{2-} ions. Use of Ca salts have a distinct advantage over those of Na or Na, as Ca prevents deflocculating in heavy textured soils and leads to easy filtration. SO_4^{2-} in the extract can be estimated turbid metrically using a spectrophotometer. Sulphur in the extract can also be measured by colorimetric method using barium chromate but the turbimertic method given below is mainly used in the soil testing laboratories.

Apparatus Required

Conical flask, volumetric flask, pipette, balance, mechanical shaker, measuring cylinder, beaker, funnel and spectrophotometer etc.

Reagents required

a. 0.15% calcium chloride solution: Dissolve 1.5 g of calcium chloride dehydrate in about 500 mL of distilled water and make up volume to 1 L.

b. Barium chloride crystal: Grind barium chloride crystals to pass through a 30 mesh sieve and retain on a 60 mesh sieve, and store in a clean bottle.

c. Standard sulphur solution: Dissolve 0.5434g of oven dried AR grade K_2SO_4 in distilled water and dilute to 1 L. This contains 100 ppm sulphur.

d. Gum acacia solution: Dissolve 0.5 g of gum acacia in distilled water and dilute to 100 mL.

Preparation of standard curve

i. Take 0.25,0.50,1.0,2.5 and 5.0 mL of standard S solution in 25 mL volumetric flask and add 10 mL of extracting solution(0.15% calcium chloride solution) to each flask. For blank take 10 mL extracting solution in a 25 mL volumetric flask.

ii. Add 1 g $BaCl_2$ crystals to each flask and swirl to dissolve the crystals.

iii. Add 1 mL of 0.25% gum acacia solution, make up the volume with distilled water and shake manually.

iv. Within 5-30 minutes of development of turbidity (white colour), read the absorbance at 420 nm on a spectrophotometer or on a colorimeter using blue filter.

v. Draw a curve taking S concentration on X axis and absorbance reading on Y axis.

Procedure

- Weight 10 g air dry soil in a 150 mL conical flask and add 50 mL of 0.15% $CaCl_2$ solution.
- Shake 30 minutes on a rotary shaker and filter through filter paper.
- Take 10 mL of the filtrate in a volumetric flask, follow the steps given for standard curve to develop the turbidity and record the absorbance.

Calculations

a. Weight of the soil = 10 g

b. Volume of extract=50 ml

c. First dilution=50/10=5 times

d. Volume of aliquot taken=10

e. Final volume make up=25

f. Second dilution=25/10=2.5times

g. Total dilution=5x2.5=12.5 times

h. Conc. from standard curve= R

i. Available S in soil (ppm)=Rx dilution factor

j. Available S in soil (kg/ha)=ppm*dilution factor*2.24

Interpretation

Available soil S (kg ha^{-1})	Soil rating
<22.4	Low
22.4-35	Medium
>35	High

13

Determination of Extractable Calcium Carbonate in Soil

Carbonate is a natural constituent of many soils occurring as sparingly soluble, alkaline earth carbonate, chiefly as calcite and dolomite. Most carbonate minerals found in local soils are calcite and account 90% of total soil carbonates. The determination of total carbonate such as $CaCO_3$in soil is of great interest on account of high useful ess for diagnosing soil status in terms of structure, texture, biological activity or nutrient content. In calcareous soils carbonates exert a major influence on the chemical properties. Also carbonate content is used as a differentiating criterion for some classes at the family level of soil taxonomy.

Principle

A qualitative rapid test for free calcium carbonate is made by adding 10 mL of diluted HCl(1:3) to about 5 g of air dry soil on a watch glass and observing the intensity of effervescence which results from the evolution of CO Based on the degree of effervescence the amount of calcium carbonate present in the soil can be broadly categorized into "traces or negligible, low, medium, high and very high". Soils having traces to low amounts of calcium carbonate do not pose any problem. But those showing medium to very high degree of effervescence, a separate quantitative analysis should be carried out using Puri's rapid titration method.

Puri's Rapid Titration Method

$$CaCO_3 + 2\ HCl \longrightarrow CaCl_2 + H_2O + CO_2$$

Apparatus Required

Burette, pipette, conical flask and mechanical shaker etc.

Reagent

a. 1N Hydrochloric acid

b. 1N NaOH

c. Calcium sulphate powder

d. Bromothymol blue indicator: Grind 1 g bromothymol blue in an agate pestle-morter and dissolve in 100 mL ethanol.

Procedure

- Weigh 2.5 g air dry soil and transfer to a 150 mL conical flask, Add 50 mL of 1 N HCl.
- Stir vigorously for 1 hour and allow to settling.
- Pipette out 10 mL of the supernatant liquid
- Transfer in a 100 mL conical flask
- Add 3-4 drops of bromothymol blue indicator.
- Titrate with 1 N NaOH solution.
- The colour of indicator fades as the end point.
- Note the reading (mL)
- Carry out a blank in same way.

Calculation

$CaCO_3$(%) in soil=(Blank titration-Actual titration) × wt. of soil sample(g)

14

Determination of Extractable Calcium and Magnesium in Soil

Calcium (Ca) and magnesium (Mg) are the two most abundant alkaline earth cations in soils. They both occur as soluble ions in soil solution, exchangeable, non-exchangeable and in organic material. Soils from arid regions are richer in Ca (1-2% CaO), while those from humid zones have a low content (0.01-0.02% CaO). On the other hand exchangeable Mg is found in very small quantities in calcareous soils. Sometimes, in acid soils, Mg^{2+} represents the largest quantity of the sum of exchangeable bases.

Principle

The method is based on the principle that calcium, magnesium and a number of other cations form stable complexes with versenate (ethylendiaminetetraacetic acid disodium salt, EDTA) at different pH. A known volume of standard calcium solution is titrated with standard 0.01N EDTA solution using murexide (ammonium purpurate) indicator in the presence of NaOH solution.

The end point is a change of color from orange red to purple at pH 12 when the whole of Ca forms a complex with EDTA. Calcium plus Mg are also present in the solution and can be titrated with 0.01N EDTA using buffer solution and a few drops Eriochrome Black T indicator. The end point is a change of color from red to blue at pH 10. Beyond pH 10, magnesium is not bound strongly to Erichrome Black-T indicator to give a distinct end point.

Apparatus Required

Burette, pipette, conical flask and mechanical shaker etc.

Reagents

a. Buffer Solution (NH_4Cl-NH_4OH): Dissolve 67.5 g NH_4Cl in 570 mL concentrated NH_4OH, and transfer the solution to a 1-L flask, let it cool, and bring to volume.

b. Eriochrome Black Indicator: Dissolve 0.5 g eriochrome black with 4.5 g hydroxylamine hydrochloride in 100 mL 95% ethyl alcohol. Prepare a fresh batch every month.

c. Ethylene Diaminetetraacetic Acid Solution (EDTA), 0.01 N: Dissolve 2 g EDTA, and 0.05 g magnesium chloride ($MgCl_2$) in DI water, and bring to 1-L volume.

d. Sodium Hydroxide Solution (NaOH), 2 N: Dissolve 80 g NaOH in about 800 mL DI water, transfer the solution to a 1-L flask, cool, and bring to volume.

e. Ammonium Purpurate Indicator ($C_8H_8N_6O_6$) (Murexide): Mix 0.5 g ammonium purpurate (Murexid) with 100 g potassium sulfate (K_2SO_4).

f. Ammonium Acetate Solution (NH_4OAc), 1 N: Add 57 mL concentrated acetic acid (CH_3COOH) solution to 800 mL DI water, and then add 68 mL concentrated ammonium hydroxide (NH_4OH) solution, mix well, and let the mixture cool.Adjust to pH 7.0 by adding more CH_3COOH or NH_4OH, and bring to 1-L volume with DI water.

Procedure

A. Extraction

- Weigh 10 g air-dry soil (< 2-mm) into a 250-mL flask.
- Add 50 mL 1 N NH_4OAcsolution (ratio 1:5).
- Shake for 30 minutes on a reciprocate shaker at 200-300 rpm.
- Filter suspension using a Whatman No.1 filter paper to exclude any soil particles, and bring the extract to a 50-mL volume with 1 N NH_4OAcsolution.

B. Measurement

Calcium

a. Pipette 10-20 mL soil extract, into a 250-mL Erlenmeyer flask.

b. Dilute to 20-30 mL with DI water, add 2-3 mL 2 N NaOH solution.

c. Add about 50 mg ammonium purpurate (Murexide) indicator.

d. Titrate with 0.01 N EDTA until the color changes from red to lavender or purple. Near the end point, EDTA should be added one drop every 10 second since the color change is not instantaneous.

e. Always run a blank containing all reagents but no soil, and treat it in exactly the same way as the samples; and subtract the blank titration reading from the readings for all samples.

Calcium plus Magnesium

a. Pipette 10 – 20 mL soil extract into a 250-mL Erlenmeyer flask.

b. Dilute to 20 – 30 mL with DI water, add 3-5 mL buffer solution.

c. Add a few drops eriochrome black indicator.

d. Titrate with 0.01 N EDTA until the color changes from red to blue.

e. Always run a blank containing all reagents but no soil, and treat it in exactly the same way as the samples; and subtract the blank titration reading from the readings for all samples.

Observations and Calculations

a. Weight of soil taken=W g

b. Volume of extract made=V mL

c. Volume of aliquot taken for analysis=V1 mL

d. Volume of EDTA used for the sample=V2 mL

e. Volume of EDTA used for Blank=V3 mL

f. Normality (N) of EDTA=0.01

Hence, meq of (Ca+Mg)/100 g of soil= (V2-V3)x0.01xV/V1x100/W

meq of Ca/100 g of soil= (V2-V3)x0.01xV/V1x100/W

Now meq of Mg/100 g soil= meq of (Ca+Mg)/100 g of soil - meq of Ca/100 g of soil

15

Determination of Carbonate and Bicarbonate in Soil

Principle

Carbonate (CO_3) and bicarbonate(HCO_3) ions are species of the same acid, carbonic acid. Their proportionate content is a function of pH. The CO_3 starts to form as pH rises above 8.4. Carbonate and bicarbonate are generally determined in soil saturation extract by titration with 0.01 N H_2SO_4 to pH 8.3 and 4.5, respectively.

Apparatus Required

Burette, pipette, conical flask and mechanical shaker etc

Reagents

a. Methyl Orange Indicator [4-$NaOSO_2C_6H_4N:NC_6H_4$/-4-N $(CH_3)_2$], (F.W. 327.34), 0.1% : Dissolve 0.1 g methyl orange indicator in distilled water, and bring to 100-mL volume.

b. Sulfuric Acid Solution (H_2SO_4), 0.01 N:Add 28 mL concentrated $H_2SO_4$to about 600 – 800 mL DI water in a 1-L flask, mix well, let it cool, and bring to 1-L volume. This solution contains 1 N H_2SO_4 solution (Stock Solution). Pipette 10 mL Stock Solution to 1-L flask, and bring to volume with distilled water. This solution contains 0.01 N H_2SO_4.

c. Phenolphthalein Indicator, 1%: Dissolve 1 g phenolphthalein indicator in 100 mL ethanol.

Procedure

A. Extraction

- Soluble carbonate and bicarbonate can be obtained in a water extract from a saturated paste as for pH and EC determinations.

- Weight 20 g soil in a conical flak.
- Add 100 mL of distilled water and place the flask in a shaking machine for one hour for equilibration.
- Filter the suspension. The filtrate is used for different estimations.

B. Measurement

- Pipette 10 – 15 mL soil saturation extract in a wide-mouthed porcelain crucible or a 150-mL conical flask.
- Add 1 drop phenolphthalein indicator. If pink color develops, add 0.01 *N* $H_2SO_4$by a burette, drop by drop, until the color disappears.
- Take the reading, *y*.
- Continue the titration with 0.01 *N* H_2SO_4after adding 2 drops 0.1 % methyl orange indicator until the color turns to orange.
- Take the reading, t.
- Always run two blanks containing all reagents but no soil, and treat them in exactly the same way as the samples. Subtract the blank titration reading from the readings for all samples.

Calculations

Soluble CO_3 (meq/L)=2y×N×1000/V

Soluble HCO_3 (meq/L)=(t-2y)×N×1000/V

Where:

2 = Valance of carbonate

Y = Volume of titrant against phenolphthalein indicator (mL)

t = Volume of titrant against methyl orange indicator (mL)

V = Volume of soil extract used for titration (mL)

N = Normality of H_2SO_4 solution

16

Determination of Available Chloride in Soil

Soluble chloride (Cl) concentration in the extract is determined by silver nitrate titration. This method quantifies the concentration of Cl (meq/L) in the saturation paste extract. Chloride may be determined using anion selective electrode (potentiometric). Plant tolerance to Cl can be related to its concentration in the soil saturation paste extract. The method detection limit is approximately 0.1 meq/L, dependent on the method of analysis, and is generally reproducible within ± 10 %.

Apparatus Required

Burette, pipette, conical flask and mechanical shaker etc

Reagents

a. Potassium Chromate Solution (K_2CrO_4), 5% in water:Dissolve 5 g K_2CrO_4in 50 mL DI water. Add dropwise 1 N silver nitrate (AgNO3) until a slight permanent red precipitate is formed. Filter, and bring to 100-mL volume with DI water.

b. Silver Nitrate Solution ($AgNO_3$), 0.01 N:Dry 3 - 4 g AgNO3 in an oven at 105 °C for 2 hours. Cool in a desiccator, and store in a tightly stoppered bottle. Dissolve 1.696 g dried AgNO3 in DI water, and bring to 1-L volume.

c. Sodium Chloride Solution (NaCl), 0.01 N:Dry 2-3 g NaCl in an oven at 110 °C for 3 hours. Cool in a desiccator, and store in a tightly stoppered bottle. Dissolve 0.585 g dried NaCl in DI water, and bring to 1-L volume.

Procedure

A. Extraction

- Soluble carbonate and bicarbonate can be obtained in a water extract from a saturated paste as for pH and EC determinations.

- Weight 20 g soil in a conical flak.
- Add 100 mL of distilled water and place the flask in a shaking machine for one hour for equilibration.
- Filter the suspension. The filtrate is used for different estimations.

B. Measurement

- Pipette 5-10 mL soil saturation extract in a wide-mouth Erlenmeyer flask (150-mL).
- Add 4 drops 5 % K_2CrO_4solution.
- Titrate against 0.01 N $AgNO_3$until a permanent reddish-brown color appears.
- Always run two blanks containing all reagents but no soil, and treat them in exactly the same way as for the samples. Subtract the blank titration reading from the readings for all samples.
- In order to standardize the $AgNO_3$ solution used in the determination of Cl:

a. Pipette 10 mL 0.01N NaCl solution in a wide-mouth Erlenmeyer flask.

b. Add 4 drops 5 % K_2CrO_4solution.

c. Titrate against 0.01 N AgNO3 solution, until a permanent reddish-brown color appears.

d. Take the reading, and calculate $AgNO_3$normality:

$$N_{AgNO3} = 10 \times N_{NaCl} / V_{AgNO3}$$

Where:

N_{AgNO3} = Normality of $AgNO_3$ solution

V_{AgNO} = Volume of AgNO3 solution used (mL)

N_{NaCl} = Normality of NaCl solution

Calculations

Soluble Cl (meq/L)=(V-B)×N×1000/V1

Where:

V = Volume of 0.01 N $AgNO_3$ titrated for the sample (mL)

B = Blank titration volume (mL)

V1 = Volume of extract used for titration (mL)

N = Normality of $AgNO_3$ solution

17

Determination of Micronutrients in Soil by Atomic Absorption Spectrophotometer

All atoms can absorb light at certain discrete wavelengths corresponding to the energy requirement of the particular atom. When at ground state the atom absorbs light it is transformed into the excited state. It is the same atom containing more energy. This energy is measured in relation to the ground state and a particular excited state say for example in case of Na may be 2.2 eV (electron volts) above the ground state.

Each transition between different electronic energy states is characterized by a different energy and by a different wavelength. These wavelengths are sharply defined and when a range of wavelengths is surveyed, each wavelength shows as a sharp energy maximum (a spectronic line). These characteristic lines distinguish atomic spectra. The lines, which originate in the ground state of atom, are most often of interest in atomic absorption spectroscopy. These are called the resonance lines. The atomic spectrum, characteristic of each element, then comprises a number of discrete lines, some of which are resonance lines. Most of the other lines arise from excited states rather than the ground state. The lines of excited states are not useful generally in atomic absorption analysis as most of the atoms in a practical atomizer are found in the ground state.

The relationship of light absorbed by the atom in ground state and their concentration in the solution is defined in the fundamental laws of light absorptions.

Lambert's Law: The portion of lightabsorption by a transparent medium is independent of the intensity of the incidence light and each successive unit thickness of the medium absorbs an equal fraction of the light passing through it.

Beer's Law: Light absorption is proportional to the number of absorbing atoms in the sample.

The combined Beer - Lambert law may be given as :

$I_t = I_o - (abc)$

thus log10 $\frac{I_0}{I_t}$ = abc = Absorbance

Where, I_0= incident radiation power

I_t=Transmitted radiation power

a= absorption coefficient

b=length of absorption path

c=concentration of absorbing atoms

i.e. the absorbance is proportional to the concentration of the elements for a given absorption path length at any given wave length.

In principle, it might be possible to calculate the concentration directly from the above equation. In practice, however, the a and b are constants hence the variation of results is directly related the concentration of atoms. For analysis, the absorbance of different concentration of standard solution is first measured with the help of atomic absorption spectrophotometer and then the results of unknown samples are compared with the standards and thus concentration of unknown sample is calculated.

Atomic absorption spectrophotometer

Atomic absorption spectro-photo-meter is based on the principle that when atomic vapours of an element are irradiated by the radiation of a characteristic wavelength (i.e. the light from a source whose emission lines are those of the element in question), they absorb in direct proportion to the concentration of the element being determined.

Instrument features

A wide range of atomic absorption spectrophotometer is available today, all of them have the basic features in common and consist of the following components:

(a) A Light source

A Light source emits the spectrum of the element to be determined. The most widely used light source is hollow cathode lamp which is designed and operated in such a way that the lines to be measured are sharp, of stable intensity and free from background.

(b) Atomizer-Burner assembly

A means of producing atomic vapours of the element to be analyzed. The solution to be analyzed is drawn by capillary and converted into stream of compressed air to a fine spray which after condensation of larger droplets is mixed with the fuel gas acetylene and burnt in a long flame (at 2100-2400°C) in a stainless steel burner.

(c) A Monochromator

It isolates the absorbing resonance lines from other non absorbing lines. When the light coming from the HCL, after traversing the flam, enters the monochromator which is already set at the wavelength of the resonance lines of the desired element, the monochromator performs its function.

(d) A Detector

It measures the magnitude of absorption of the characteristic radiation.

(e) A Photomultiplier Tube

It amplifies the absorption signal and converts the light radiation into electrical energy.

(f) A readout system

It measures the absorbance in volts. It is normally a strip chart recorder, a digital display, a meter or printer. The presently available AAS have features like automatic calibration with one or more standards, automatic curve corrections, automatic and foolproof gas switching and calculation of average and standard deviations in repetitive runs

DTPA Method

The diethylene triamine pentaacetic Acid(DTPA) test of Lindsay and Norvell (1978) is commonly used for evaluating fertility status with respect to micronutrient cations, i.e., Fe, Zn, Mn, and Cu. The DTPA method is an important and widely used chelating agent, which combines with free metal ions in the solution to form soluble complexes of elements. The DTPA method has a capacity to complex each of the micronutrient cations as 10 times of its atomic weight. The capacity ranges from 550 to 650 mg/kg depending upon the micronutrient cations. However, the universal soil test for alkaline soils (i.e., AB-DTPA) is equally effective for determining micronutrient cations in alkaline soils. Deficiencies of Mo, Cl, Ni and Co are not known to occur in alkaline soils.

Apparatus Reqiured

a. Atomic absorption spectrophotometer

b. Mechanical shaker, reciprocal

c. Volumetric flask

Reagents

DTPA Extraction Solution

- Weigh 1.97 g diethylene triamine pentaacetic acid (DTPA), and 1.1 g calcium chloride (CaC12) or [(1.47 g calcium chloride dihydrate ($CaCl_2.2H_2O$)] into a beaker. Dissolve with DI water and then transfer to a 1-L volume.
- Into another beaker, weigh 14.92 g (or add 13.38 mL) Triethanolamine (TEA), transfer with DI water into the 1-L flask, and then bring to about 900-mL volume.
- Adjust the pH to exactly 7.3 with 6N hydrochloric acid (HCl), and bring to 1-L volume. This solution contains 0.005 M DTPA, 0.1 M TEA, 0.1 M $CaCl_2$.

Stock Standard Solutions: Thestandard solutions of different micro-nutrients should preferably be prepared by using their wires. Dissolve 1g wire in a minimum volume of 1:1 nitric acid and dilute to 1000ml with distilled water to obtain 1000 µg/ml solution of micro-nutrient, or take salts of metals as follows:

a. Zn- 4.398g l^{-1} $ZnSO_4.7H_2O$

b. Cu- 3.929g l^{-1} $CuSO_4.5H_2O$

c. Fe- 4.977g l^{-1} $FeSO_4.7H_2O$

d. Mn- 3.598g l^{-1} $MnSO_4.H_2O$

Standard Stock Solutions

Prepare a series of Standard Solutions for micronutrients in DTPA extraction solution:

1. Iron (Fe) standard solution

- Pipette 10 mL Fe Stock Solution (1000 ppm) in 100 mL flask and then dilute to volume with DTPA solution. This solution contains 100 ppm Fe (Diluted Stock solution).

- Pipette 1, 2, 3, 4 and 5 mL Diluted Stock Solution to 100 mL numbered flask and then dilute to volume DTPA solution. These solutions contain 1, 2, 3, 4, and 5 ppm Fe, respectively.

2. Zinc (Zn) standard solution

- Pipette 10 mL Zn Stock Solution (1000 ppm) in 100 mL flask, and then dilute to volume with DTPA solution. This solution contains 100 ppm Zn (Diluted Stock Solution).
- Pipette 10 mL Diluted Stock Solution to 100-mL flask, and then dilute to volume DTPA solution. This solution contains 10 ppm Zn (Second Diluted Stock Solution).
- Pipette 1, 2, 4, 6, 8 and 10 mL Second Diluted Stock Solution in 50-mL numbered flasks, and then dilute to volume DTPA solution. These solutions contain 0.2, 0.4, 0.8, 1.2, 1.6 and 2.0 ppm Zn, respectively.

3. Copper (Cu) standard solution

- Pipette 10 mL Cu Stock Solution (1000 ppm) in 100-mL flask, and then dilute to volume with DTPA solution. This solution contains 100 ppm Cu (Diluted Stock Solution).
- Pipette 10 mL Diluted Stock Solution to 100-mL flask, and then dilute to volume DTPA solution. This solution contains 10 ppm Cu (Second Diluted Stock Solution).
- Pipette 2, 3, 4, 5, 6 and 7 mL Second Diluted Stock Solution in 50-mL numbered flasks, and then dilute to volume DTPA solution. These solutions contain 0.4, 0.6, 0.8, 1.0, 1.2 and 1.4 ppm Cu, respectively.

4. Manganese (Mn) standard solution

- Pipette 10 mL Mn Stock Solution (1000 ppm) in 100-mL flask and then dilute to volume with DTPA solution. This solution contains 100 ppm Mn (Diluted Stock Solution).
- Pipette 10 mL Diluted Stock Solution to 100-mL flask, and then dilute to volume DTPA solution. This solution contains 10 ppm Mn (Second Diluted Stock Solution).
- Pipette 2, 3, 4, 5, 6 and 7 mL Second Diluted Stock Solution in 50-mL numbered flasks, and then dilute to volume DTPA solution. These solutions contain 0.4, 0.6, 0.8, 1.0, 1.2 and 1.4 ppm Mn, respectively.

Procedure

A. Extraction

- Weigh 10 g air-dry soil (2-mm) into a 150mL conical flask.
- Add 20 mL extraction solution.
- Shake for 2 hours on a reciprocal shaker.
- Filter the suspension through a Whatman No. 42 filter paper.
- Prepare atleast 10 standards using DTPA as the matrix for each element with a range 0 to 3 ppm for Zn and Cu and from 0 to 20 ppm for Fe and Mn.
- First take the reading for the standards, then, measure the elements from the filtrate by AAS.
- A blank solution containing all reagents excepting the soil, should be run to correct for contamination.

Observations and calculations

a. Weight of soil=10g

b. Volume of DTPA extract=20 mL

c. Concentration (ppm) of metal in extract as read from AAS=C

d. Concentration (ppm) of metal in blank as read from AAS=C_b

e. Dilution factor=20/10

f. DTPA extractable metal in the soil(ppm)=$(C- C_b)x2$

g. DTPA extractable metal in the soil(kg/ha)=$(C- C_b)x2x2.24$

18

Determination of Available Boron in Soil

Boron occurs as anion in soils and isrequired by plants in very small quantity. Water soluble B makes the estimate of its availability to plants. Total boron in soils varies from 20 to 200 mg kg^{-1} and available (water soluble) boron in soils ranges from 0.03 to 12 mg kg^{-1} respectively. The threshold value ranging from 0.1 to 0.5 mg kg^{-1} (water soluble B) depends upon the soil type, crops, and other factors, below which the response to applied boron may be expected. Its availability is affected by soil pH as under:

Deficiency of B is generally observed in old acid leached soils.

- Availability increased with the rise in soil pH having significant positive correlation with pH rising from 4.7 to 6.7.
- In neutral, saline and calcareous soils the B availability again decreases with the rise in soil pH having significant negative correlation with the rise in pH from 7.1 to 8.1. In calcareous soils B fixation occurs with the condensation of borate radical into long chains in the presence of Ca.
- In alkaline soils the availability of B is high and may be even toxic for plant growth.

Besides this the low moisture availability also causes B deficiency. Irrigation water containing Boron between 0.3 to 0.6 mg kg^{-1} can be used safely, whereas, irrigating soils with water containing 1 to 3 mg kg^{-1} B causes toxicity of B in plants.

Boron determination (Azomethine H Method)

Azomethine H forms coloured complex with H_3BO_3 in aqueous media. Over a concentration range of 0.5 to 10 μg B/ml the complex is stable at pH 5.1. Maximum absorbance occur at 420 nm with little or no interference from a wide variety of salts. This technique is rapid, reliable and more convenient to use than traditional procedures employing carmin, curcumin or quinalizarin (John *et al.*, 1975).

Apparatus

a. Spectrophotometer

b. Poly-propylene tubes 10 ml capacity.

c. Hot plate

d. Refrigerator

Reagents

a. Distilled water

b. Buffer solution: Dissolve 250 g of ammonium acetate (NH_4OAc) and 15 g of ethylenediaminetetracetic acid (EDTA disodium salt) in 400 ml of distilled water. Slowly add 125 ml of glacial acetic acid and mix.

c. Azomethine H reagent: Dissolve 0.45 g of azomethine H in 100 ml of 1% L ascorbic acid solution. Fresh reagent should be prepared weekly and stored in a refrigerator.

d. Calcium chloride 0.01M: Dissolve 1.11g of anhydrous $CaCl_2$ in 900 ml distilled water and make up the volume to 1 litre.

e. Boron standard solution: Dissolve 0.114g of Boric acid (H_3BO_3) in distilled water and adjust the volume to 1000 ml. Each ml contains 20 µg B. Dilute 5, 10, 20, 30, 40 and 50ml of the stock solution to 100 ml with distilled water to have solution with B concentration of1, 2,4,6,8 and 10 µg of B/ml respectively. Include a distilled water sample for the 0.0 µg of B/ml standard solution.

Procedure

- Weight 20 g of air dry soil sample in a 250 mL quartz and other boron free conical flask and add 40 mL 0.01 M $CaCl_2$ solution.
- Add 0.5 g of activated charcoal and boil for 5 minutes on a hot plate, filter immediately through Whatman No. 42 filter paper.
- Cool the contents to room temperature and transfer 1 mL aliquot of blank (filtrate obtained after boiling and activated charcoal), diluted B standard or filtrate into a 10 or 15 mL plastic tubes.
- Add 2 mL of buffer solution and mix.
- Add 2 ml of Azomethane H reagent, mix and after 30 minutes read the absorbance at 420 nm on a spectrophotometer.
- Prepare a standard curve plotting B concentration on X axis and absorbance on Y axis.

- Refer the absorbance reading of sample to the standard curve to obtain B content of aliquots.

Calculation

Available boron (mg/kg or ppm) in soil=Ax2

A=B content obtained from standard curve

Section II: Pesticide Residue Analysis

19

Pesticide Residue Analysis: Overview

A pesticides residue is any substance or mixture of substances in food, agricultural commodities or animal feed resulting from the use of pesticides and includes any specified derivatives such as degradation products and impurities that are considered to be of toxicological significance.

Pesticide residue analysis (PRA) may be defined as the quantitative and qualitative analyses of the representative samples drawn from agricultural field, market and environment for pesticides and their toxic metabolites.

Objectives of Pesticide Residue Analysis

- To study the persistence of pesticide in or on the soil, plant and water.
- To establish pre-harvest intervals (PHI) on the basis of multilocation trials.
- Dietary risk assessment
- To establish Maximum Residue Limits (MRLs) of pesticides.
- To monitor environmental samples.
- To know pesticide use pattern in farmer fields.

Involved steps in Pesticide Residue Analysis

Five most important steps

1. Sampling – collection, transport and storage of samples.
2. Preparation of samples
3. Extraction
4. Clean-up
5. Identification, quantification and confirmation

Reagents

a. *Florisil*: Synthetic magnesium silicate, used as adsorbent.

b. *Sodium sulphate/sodium chloride*: to remove phthalates esters.

c. *Glass wool*: Pyrex glass wool can have contaminants that interfere with determination.

d. *Celite 545*:

e. *Charcoal*: to absorb color impurities

f. *Magnesium oxide*:

Solvents

Hexane, chloroform, toluene, isopropanol, methanol, dichloromethane, ethyl acetate, isooctane, acetone, diethyl ether, acetonitrile.

Preparation of pesticide standard solutions

Organic solvents of high purity are required for the preparation of standard solutions because any interference due to impurities in solvents will lead to errors. Stock solutions and intermediate concentrations are prepared in acetone or hexane, the working standard are preferably prepared in n-hexane. And standard solutions can be prepared in any of the solvents mention above.

Certified reference standards

Reference standards should be with certified purity by authorized agency.

Primary stock solutions

For primary stock solutions the 200 ppm concentration has been recommended for all pesticides. 10 mg certified reference material should be dissolved in minimum amount of acetone and make up volume to ml with toluene in an A grade volumetric flask accurately.

Working standard solutions

Stock solutions are normally diluted 10 times to prepare intermediate working standards solutions. It may be range of 2 to 0.02 ppm as per requirement, depending on detector sensitivity and method of analysis.

2. Sampling procedure

- **Collection of primary sample**: Each primary sample should be taken from a randomly chosen position in the lot, as far as practicable.

a. Grass, forage and animal feed

i. Cut with shears at normal harvest height (usually 5 cm above the ground) the vegetation from not less than twelve areas uniformly spaced over the entire plot, leaving I meter at the edge of the plot.

ii. Crops which are harvested mechanically can be sampled by taking not less than twelve grab samples from the harvest at uniform intervals over the plot.

iii. Take 1 kg of sample for green forage and 0.5 kg for dry hay.

b. Milk

i. A total sample size of 2 litres is usually sufficient from producer or each tank.

c. Soil

i. Soil should be collected using core-type sampling tools.

ii. The soil is cleaned by removing pebbles and stones, ground and sieved through 2 mm.

iii. Systematically reduction in size by quartered to about 200-250 gm.

d. Water

i. Collect 3 liters of water represent 1 sample from 1 water resource.

- **Storage of the sample**

i. Polythene bags, glass bottles or vials may be used as containers for holding the sample.

ii. Cooling in dry ice or liquid nitrogen is the best method.

3. Sample preparation

Where the bulk sample is larger than it is required for a laboratory sample, it should be divided to provide a representative portion.

This step followed by extraction and clean-up of pesticides after that it goes to identification/quantifications/confirmation with help of chromatographic and spectroscopic technique, GLC or HPLC/ LC-MS or GC-MS.

20

Instrumentation- Gas Chromatography (GC, GC-MS & GC-MS/MS)

Chromatography is a physical method of separation in which components to be separated is distributed between two phases, one of which is stationary while other moves in a definite direction. By separating the sample into individual components, it is easier to identify (qualitative) and measure the amount (quantitative) of the various sample components. There are numerous chromatographic techniques and corresponding instruments.

Gas Liquid Chromatography (GLC) is the most widely used analytical technique for estimation of volatile and thermo stable type of pesticides with speed, accuracy, reproducibility and sensitivity.

Outline

- Moving phase is gas
- Stationary phase is liquid
- Separation occurs in vapor form
- Coupled with detection system which can detect at very low concentration.
- Operated at high temperature (100-300 ^{0}C)
- Suitable for compounds which are heat stable and form vapors.

Components of GLC

1. Gases supply & regulation: Carrier gas and flame gases; pressure regulation and flow control
2. Sample introduction/ loading: Injection port, auto sampler
3. Separation system: Oven, column; temperature control,
4. Detection system: Detectors
5. Recording system

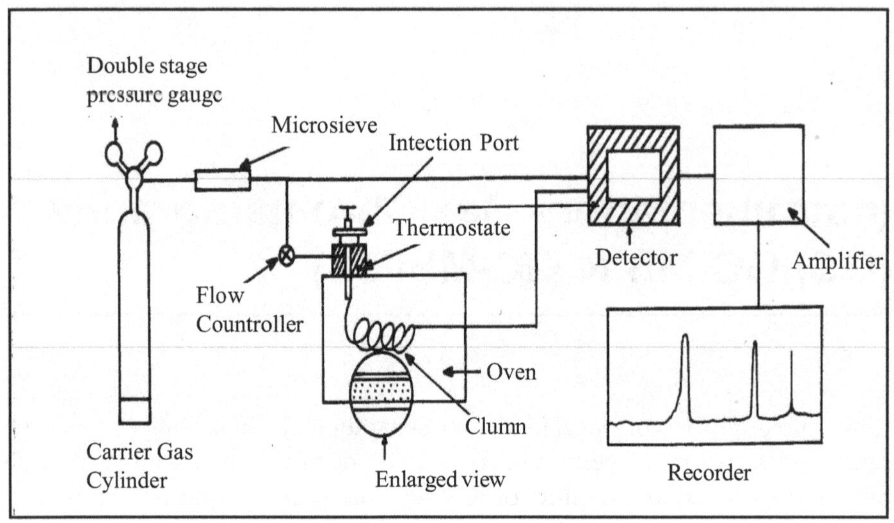

Schematic diagram of GLC

Gases supply and regulation

- Mobile Phase – Carrier Gas should be inert gas
- Best - Helium – but costly
- Most commonly used – Nitrogen gas - high purity (Grade I)
- It should not contain oxygen, moisture, organic impurities, particulate matter
- Traps used to remove traces of organic impurities, oxygen and moisture
- Flame Gases - Hydrogen and Air
- All gases are supplied in cylinders in compressed form at high pressure or gas generators.
- Pressure reduction in outlet tube using pressure regulator or gauge.
- Flow controlled using valves.
- Flow can be controlled manually or electronically (EPC or EFC)

Sample introduction/loading: (Injection port & Autosamplers)

- Sample can be introduced in liquid (solution) or vapour form.
- Samples injected into injection port using graduated microsyringe (5-10 ml) Vapour airtight syringe
- Injection - manual or auto sampler.

Columns

- Column is a main component of GC where separation occurs and this is called heart of the GC.
- Types:
  a. Packed (2-4 mm diameter)
  b. Megabore (0.53 mm diameter)
  c. Capillary (0.1-0.25 mm diameter)

Detectors used for pesticide analysis

a. Flame Ionization Detector (FID)

- Non-specific used for all organic compounds
- Form ions and collected by electrode causing flow of current
- Response a number of carbon atoms attached to C & H
- Burnt in flame (H_2 & Air)

b. Alkali Flame Ionization Detector (AFID)

- Specific for compounds.
- Burnt in presence of alkali salt to enhance ionization of compounds containing P, S and N.
- Alkali salts used are; potassium chloride, sodium sulfate, cesium bromide, rubidium chloride.

c. Flame Photometric Detector (FPD)

- Specific for P&S containing Compounds (mostly organophosphate pesticides).
- Atoms in compound excited when burnt in flame.
- Emitted light detected by photo-multiplier tube.
- Filters allow only specific light to pass, others get absorbedP: 526 & S: 394 nm.

d. Electron Capture Detector (ECD)

- Non- specific but more sensitive to halogenated compounds
- Highly sensitive

- Beta particles strike off electrons from carrier gas producing slow electrons
- Slow electrons captured by compounds in effluent

e. Mass Detector (MS & MS/MS)

- Non-specific (it can be used for all types of compounds) and mainly used for confirmation and structure elucidation.
- ***Principle***: Positive ions are deflected when passed through magnetic or electric field. The magnitude of deflection is related to mass/ charge ratio
- *Components of mass spectrometer:*
  a. Vacuum generating system
  b. Sample inlet system (interface)
  c. Ionization system
  d. Ion analyzer or separating system
  e. Ion collector or detecting system
  f. Recording system

Recording system

High speed computer is generally used to control GC-MS and recording the mass data. Recording data can be further manipulated as per requirement like comparing the recorded mass spectra with library spectra for identification of the compound.

21

Instrumentation High Performance Liquid Chromatography (HPLC)

Although Gas Liquid Chromatography (GLC) is widely used technique for pesticide residue analysis, it has some limitations. It can analyze of volatile and thermo stable type of pesticides only because it operate at high temperature. These problems have overcome in HPLC. Although HPLC can analyze both polar as well as non-polar type of compounds but it has less sensitivity than GLC.

Outline

- Moving phase is liquid.
- Stationary phase is either liquid or solid.
- Coupled with detection system which can detect at very low concentration.
- Can be used for all types of compound
- Most suitable for compounds which are polar in nature and heat labile

Components of GLC

1. Solvent delivery system
2. Sample injection system
3. Separation system: guard column, pre column and main column
4. Detection system: Detectors
5. Recording system

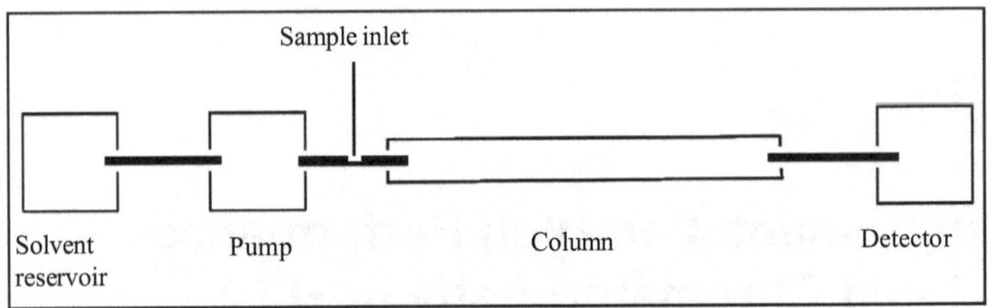

Schematic diagram of HPLC

Solvent delivery system

- The solvent delivery system consists of solvent reservoir, pumps, gradient controller and filters.
- Solvent is filtered/degassed by passing through 0.45 microns membrane.
- *Isocratic system*: when single solvent or mixture of solvent at constant proportion and flow is used.
- *Gradient system*: when proportion of constituent of mixture of solvent and flow is rate is changed with time during run.
- Most commonly used – Methanol, acetonitrile, water (HPLC grade)

Sample introduction/loading: (Injection port & Autosamplers)

- Sample can be introduced in liquidform.
- Samples injected into injection port using graduated microsyringe (5-10 ml).
- Injection - manual or auto sampler.

Separation system

- Normally operated at ambient temperature
- Fluctuation in temperature during day time
- Separation and retention time may varydue to temperature variation
- Separation system consists of guard column, pre column and main column.
- Guard column and pre column are used to protect the main column from damage.
- The separation occurs in main column.
- Length column is varies from 50 mm to 250 mm and diameter from 2 to 5 mm.

Detectors used for pesticide analysis

Various types of detectors have been used in HPLC for sensing and measuring the amount of the sample components in column effluent. Most widely used detectors are:

a. Variable wavelength UV-visible spectrometric detector(UV/VIS)

- Mostly used detector in HPLC.
- It can monitor the absorbance in UV as well as visible range.

b. Fixed wavelength absorbance detector:

- Second most commonly used detector.
- The most common used wavelength is 254 nm.

c. Multichannelphotodiode array detector (PDA)

- It has advantage over UV/VIS in a way that it monitors the absorbance simultaneously at all the wavelengths.
- This helps ascertaining the purity of given peak.
- Less sensitive and costlier than others.

d. Fluorescence detector

- Most sensitive detector.
- Compound having natural fluorescence like polynuclear aromatics, aflatoxins, aromatic amino acids, phenols, quinines etc. can be determined at very low level.
- Compounds can be reacted to produce fluorescent derivatives. common drivatising agents are densyl chloride for primary amino acids and biogenic amino acids.
- Mainly used for carbamates pesticides.

e. Refractive Index detector

- The refractive index of a medium is the ratio of the speed of light in a vacuum to the speed in the medium.
- The detector measure the change in refractive index in the eluent as the solute passes through the sample cell.
- less sensitive than UV detection

f. Mass Detector (MS, MS/MS)

- Used to obtain mass spectra characteristic of individual components of the original mixture with added advantage of micro quantitative analysis.
- Universal (Can be used for all types of compounds).
- Mainly used for confirmation and structure elucidation.

Recording system

High speed computer is generally used to control HPLC/LC-MS and recording the mass data. Recording data can be further manipulated as per requirement like comparing the recorded mass spectra with library spectra for identification of the compound.

22

Pesticide Residue Analysis in Green Fodder

Principle

Acetone is used for extraction the most nonionic pesticide residue, and residue are partitioned from acetone to dichloromethane/hexane phase. After removing traces dichloromethane, final extract is made up with acetone. This extract can be determined by GC with selective detectors. Further for organochlorine and synthetic pyrethroids extract should be cleaned up using florisil column. And for determination organophosphate extract should undergo to charcoal-celite cleanup.

For the determination carbamates, the extract is subjected to C-18 solid phase cartridge clean up.

Apparatus required

- High speed blender
- Bucchner funnel
- Separatory funnel
- Chromatographic column
- Vaccum rotary evaporator

Reagents

- Solvent used: Acetone, dichloromethane (DCM), acetonirile, hexane, methanol
- Sodium chloride, sodium sulphate, anhydrous, florisil, charcoal, celite 545, solid phase extraction cartridge, C-18.

Extraction

- Mix 100 g blended sample with 200 ml acetone for 2 min.
- Filter the extract by using Buchner funnel.

- Take an aliquot of 80 ml from extract to 1 litre separatory funnel.
- Partitioning with 200 ml of mixture of hexane:DCM (1:1,v/v), then lower aqueous phase is then transferred to another 1 litre separatory funnel.
- Organic phase of the first separatory funnel is dried by passing through anhydrous Na_2SO_4supported on pre washed cotton.
- To the separatory funnel having aqueous phase, add 10 ml of saturated NaCl solution and shake vigorously for 20 sec.
- To this, add 100 ml DCM shake vigorously and dry organic layer passed through same anhydrous Na_2SO_4that was used for previously. Repeat this step twice.
- After rinse anhydrous $Na_2SO_4$with about 50 ml of DCM. Concentrate the extract using vacuum evaporator.
- Adjust volume of extract to 7 ml with acetone. It can be used now for determination of organophosphate and nitrogen containing pesticide by GLC using FPD/NPD.

Clean-up: (for OPs, OCs, SPs)

- Run a chromatographic glass column (22 mm i.d.), add 50 ml of hexane and then pour slowly 4 g of activated florisil, followed by 2 g of Na_2SO_4.
- One ml of the extract is diluted to 10 ml with 10 % acetone in hexane.
- Transfer the solution to florisil column. Elute column at about 5 ml/min with 50 ml eluent (50 % DCM: 1.5% acetonitrile: 48.5 % hexane v/v/v).
- Concentrate florisil elute to 1 ml. the extract is ready for organochlorine and synthetic pyrethroid residue determination by GLC-ECD.
- To determine organophosphate, two ml of extract is cleaned up with celite 545 using 6 g adsorbent mixture (1:4 w/w charcoal/celite 545) in column saturated with DCM.
- Elute with 200 ml 2:1 acetone: DCM (v/v).
- Concentrate the extract and adjust the volume to definite amount by acetone.
- GC-MS can be used to analysis.

Calculations

Calculate equivalent sample weight in final solution by following formula:

$$\frac{\text{mg sample weight equivalent}}{\mu\text{l final extract (i.e. 7 ml)}} = 100 \times \frac{80}{200 + \text{W-10}} \times \frac{1}{\text{ml final vol.}}$$

Where, 100 g sample analyzed, 80 ml filtered extract filtered extract taken for hydromatrix partition, W, amount of water present in 100 g of sample, 200 ml acetone for blending, 10, adjustment for water/acetone volume concentration.

Clean-up: (for carbamates)

- Concentrate the extract obtained after L-L-P to near dryness undercurrent of nitrogen.
- Prewet C-18 cartridage (2.8 ml) with methanol and discard solvent.
- Dissolve residue with 2 ml methanol and transfer quantitatively onto prewet C-a8 cartridge.
- Elute cartridge with additional methanol until collected volume is about 5 ml. add methanol to make the volume 5 ml.
- Cleaned up extract contain concentration of sample with same formula.
- Determined N-methyl carbamates with HPLC.

23

Pesticide Residue in Dry Fodder

Principle: Acetone is used for extraction the most nonionic pesticide residue, and residue are partitioned from acetone to dichloromethane/hexane phase. After removing traces dichloromethane, final extract is made up with acetone. This extract can be determined by GC with selective detectors FPD-P/NPD without cleanup. Further for organochlorine and synthetic pyrethroids extract should be cleaned up using florisil column. And for determination organophosphate extract should undergo to charcoal-celite cleanup.

Apparatus required

- High speed blender
- Bucchner funnel
- Separatory funnel
- Chromatographic column
- Vaccum rotary evaporator

Reagents

- Solvent used: Acetone, dichloromethane (DCM), acetonirile, hexane, methanol
- Sodium chloride, sodium sulphate, anhydrous, florisil, charcoal, celite 545, solid phase extraction cartridge, C-18.

Extraction

- Soak 15-25 g chopped/ blended sample with 300 ml acetone/water 65:35 for 2 hrs.
- Blend for 2 min at high speed.
- Filter the extract by using Buchner funnel
- Take an aliquot of 80 ml from extract to 1 litre separatory funnel

- Partitioning with 200 ml of mixture of hexane: DCM (1:1,v/v), then lower aqueous phase is then transferred to another 1 litre separatory funnel.
- Organic phase of the first separatory funnel is dried by passing through anhydrous Na_2SO_4supported on pre washed cotton.
- To the separatory funnel having aqueous phase, add 10 ml of saturated NaCl solution and shake vigorously for 20 sec.
- To this, add 100 ml DCM shake vigorously and dry organic layer passed through same anhydrous Na_2SO_4that was used for previously. Repeat this step twice.
- After rinse anhydrous $Na_2SO_4$with about 50 ml of DCM. Concentrate the extract using vacuum evaporator.
- Adjust volume of extract to suitable definite volume with acetone. It can be used now for determination of organophosphate and nitrogen containing pesticide by GLC using TID/FPD/NPD.

Clean-up

- To run a chromatographic glass column (22 mm i.d.), add 4 g of activated florisil, followed by 2 cm layer of Na_2SO_4.
- Pre wet column with 15 ml of hexane and do not allow column to go dry.
- One ml of the extract is diluted to 10 ml with 10 % acetone in hexane.
- Transfer the solution to florisil column, rinse container with 2x3 ml portion of hexane.
- Elute column at about 5 ml/min with 50 ml eluent (50 % DCM: 1.5% acetonitrile: 48.5 % hexane v/v/v).
- Concentrate elute to 1 ml. the extract is suitable for organochlorine and synthetic pyrethroid residue determination by GLC-ECD.

Clean-up: (for carbamates)

- Concentrate the extract obtained after L-L-P to near dryness undercurrent of nitrogen.
- Prewet C-18 cartridage (2.8 ml) with methanol and discard solvent.
- Dissolve residue with 2 ml methanol and transfer quantitatively onto prewet C-a8 cartridge.
- Elute cartridge with additional methanol until collected volume is about 5 ml. add methanol to make the volume 5 ml.

- Cleaned up extract contain concentration of sample with same formula.
- Determined N-methyl carbamates with HPLC.

Calculations

Calculate equivalent sample weight in final solution by following formula:

$$\frac{mg\ sample\ weight\ equivalent}{\mu l\ final\ extract\left(\text{i.e. 7 ml}\right)} = 100 \times \frac{80}{200 + \text{W-10}} \times \frac{1}{\text{ml final vol.}}$$

Where, 100 g sample analyzed, 80 ml filtered extract filtered extract taken for hydromatrix partition, W, amount of water present in 100 g of sample, 200 ml acetone for blending, 10, adjustment for water/acetone volume concentration.

24

Pesticide Residue Analysis in Soil

Principle

Pesticide residues are extracted from soil using a mixture of acetone/water in a soxhlet apparatus for 7-8 hours. This extract can be determined directly by GLC.

Scope: This method is suitable for most OPs, OCs, SPs, and herbicides residue analysis.

Apparatus required

- Soxhlet apparatus
- Separatory funnel
- Chromatographic column
- Vaccum rotary evaporator

Reagents

- Solvent used: Acetone, hexane, methanol
- Liquid ammonia

Procedure

- Take 100 gm of soil. Add few drops of liquid ammonia. Mix well and leave for half and hour till ammonia get evaporated.
- Transfer contents in into Soxhlet apparatus.
- Reflux using hexane: acetone (1:1 v/v) for 6-8 hours.
- Concentrate the extract using vacuum evaporator.
- Adjust volume of extract and make up thefinal volume to 10 ml with acetone: hexane (10:90).
- It can be used now for determination of pesticide by GLC.

However, in case of sample having impurities or any contaminations, adopt suitable clean-up method.

Calculations

Calculate equivalent sample weight in final solution by following formula:

$$\frac{mg\ sample\ weight\ equivalent}{\mu l\ final\ extract\left(i.e.\ 7\ ml\right)} = 100 \times \frac{80}{200 + W - 10} \times \frac{1}{ml\ final\ vol.}$$

Where, 100 g sample analyzed, 80 ml filtered extract filtered extract taken for hydromatrix partition, W, amount of water present in 100 g of sample, 200 ml acetone for blending, 10, adjustment for water/acetone volume concentration.

25

Pesticide Residue Analysis in Milk

Principle

Pesticide residues in milk are derived from the intake of pesticide contaminated feed and fodder.

Scope

This method is suitable for most OPs, OCs, SPs, and herbicides residue analysis.

Apparatus required

- Homogenizer
- Centrifuge
- Nitrogen evaporator

Reagents

- Anhydrous magnesium sulfate (MgSO4): Powder form; purity >98%; heated in bulk to 500°C for >5 h to remove phthalates and residual water.
- Acetonitrile (MeCN): Quality of sufficient purity that is free of interfering compounds.
- Acetic acid (HOAc): Glacial; quality of sufficient purity that is free of interfering compounds.
- 1% HOAc in MeCN: Prepared on a v/v basis (e.g., 10 mL glacial HOAc in a 1 L MeCN solution).
- Anhydrous sodium acetate (NaOAc): Powder form (NaOAc ·3H2O may be substituted, but 0.17 g per g sample must be used rather than 0.1 g anhydrous NaOAc per g sample).
- Primary secondary amine (PSA) sorbent: 40 mm particle size (Varian Part No. 12213024 or equivalent). (Note: Premade dispersive-SPE tubes are now available from at least 3 vendors.)
- C18 sorbent (optional): 40 mm particle size, if samples contain >1% fat.

Extraction and Clean up by QuEChERS

The QuEChERS (quick, easy, cheap, effective, rugged, and safe) method uses a single-step buffered acetonitrile (MeCN) extraction and salting out liquid–liquid partitioning from the water in the sample with $MgSO_4$. Dispersive-solid-phase extraction (dispersive-SPE) cleanup is done to remove organic acids, excess water, and other components with a combination of primary secondary amine (PSA) sorbent and MgSO4; then the extracts are analyzed by mass spectrometry (MS) techniques after a chromatographic analytical separation.

Procedure

- Take 7.5 ml milk sample in a clean 50 ml capacity propylene centrifuge tube.
- Add 7.5 ml of LC-MS grade water and mix thoroughly
- Keep in deep freezer at 0°C to -10 °C for 20 min.
- Add 15 ml of 1% acetic acid in acetonitrile (v/v) &10 ml of distilled water.
- Add 6 g $MgSO_4$ (anhydrous) and 1.5 g sodium chloride (anhydrous)
- Shake vigorously for 1 min followed by vortex mixing
- Centrifuge for 3 min at 2000 rpm
- Transfer 6 ml supernatant to 15 ml polypropylene centrifuge tube containing 300 mg PSA + 900 mg $MgSO_4$ + 500 mg C-18 to the supernatant and vortex the sample for 30 sec
- Centrifuge for 1 min at 2000 rpm to get clean supernatant
- Transfer 3 ml supernatant in 15 ml capacity glass test tube.
- Evaporate the sample to dryness with N_2on turbovap at 45°C
- Make up to 1 ml with petroleum spirit: acetone (3:1 v/v) for estimation on GC analysis.

Calculations

$$\text{Residues } (\mu g/ml) = \frac{\text{Ht/Area of the sample}}{\text{Ht/Area of the std.}} \times \frac{\mu l \text{ of standard injected}}{\mu l \text{ of sample injected}} \times \frac{\text{Conc. of standard } (\mu g/ml)}{\text{weight of sample } (1.5\text{ g})} \times \text{final vol. } (1\text{ ml})$$

26

Pesticide Residue Analysis from Concentrated Animal Feed

Scope

This method is suitable for most OPs, OCs, SPs, and herbicides residue analysis.

Apparatus required

- Homogenizer
- Centrifuge
- Nitrogen evaporator

Reagents

- Anhydrous magnesium sulfate (MgSO4): Acetonitrile (MeCN): Quality of sufficient purity that is free of interfering compounds.
- Acetic acid (HOAc):
- 1% HOAc in MeCN:
- Anhydrous sodium acetate (NaOAc):
- Primary secondary amine (PSA) sorbent:
- C18 sorbent (optional):

Extraction and Clean up by QuEChERS

The QuEChERS (quick, easy, cheap, effective, rugged, and safe) method uses a single-step buffered acetonitrile (MeCN) extraction and salting out liquid–liquid partitioning from the water in the sample with $MgSO_4$. Dispersive-solid-phase extraction (dispersive-SPE) cleanup is done to remove organic acids, excess water, and other components with a combination of primary secondary amine (PSA) sorbent and MgSO4; then the extracts are analyzed by mass spectrometry (MS) techniques after a chromatographic analytical separation.

Procedure

- Take 25 ± 0.1 g sample from coarsely grounded animal feed sample in 500 ml capacity wide mouth glass bottle containing 10 g NaCl.
- Add 50 ml acetonitrile.
- Homozenize with high speed vertical hozenizer for 3 min.
- Filter the content through pre washed cotton bed into 250 ml capacity conical flask
- Draw 8 ml filtrate in a 25 ml capacity tube containing 5 g Na_2SO_4.
- Shake vigorously and allow it to settle.
- Draw 4 ml supernatant to 15 ml polypropylene centrifuge tube containing 100 mg PSA + 600 mg $MgSO_4$
- Centrifuge for 1 min at 2500 rpm
- Take 2 ml supernatant in 10 ml capacity test tube and evaporate the sample to dryness with N_2 on turbovap at 45°C
- Make final volume to 2 ml with petroleum spirit : acetone (9:1 v/v)
- Use 1 ml for estimation of organophosphates (on PFPD/FPD/TSD/FTD)
- Use remaining 1 ml, for column cleanup (For OCs and SPs).

Column Cleanup

- Add 25 ml of distilled petroleum spirit (40-60 °C) in sintered glass column (1.5 cm i.d. and 30 cm length).
- Add 1 g Na_2SO_4, 4 g florisil and 1 g Na_2SO_4in the column from bottom to top.
- Drain the solvent leaving 1 cm above the top of the column.
- Load 1 ml (equivalent 0.5 g sample) of sample in column.
- Elute freshly prepared 50 ml mixture of organic solvents contacting 50 % DCM + 1.5 % acetonitril + 48.5 % petroleum spirit to the column.
- Collect elute in 150 ml capacity of conical flask.
- Evaporate to near dryness on vacuum evaporator at 45 °C.
- Transfer the sample in 15 ml capacity glass test tube frequent washing of mixture of acetone: petroleum spirit (1:9 v/v).

- Reduce the volume to 1 ml under gentle stream of nitrogen in TurboVap at 45 °C
- Estimate the residue on GC-ECD

Calculations

$$\text{Residues } (\mu g/ml) = \frac{\text{Ht/Area of the sample}}{\text{Ht/Area of the std.}} \times \frac{\mu l \text{ of standard injected}}{\mu l \text{ of sample injected}} \times \frac{\text{Conc. of standard } (\mu g/ml)}{\text{weight of sample } (0.5\ g)} \times \text{final vol. } (1\ ml)$$

Appendices

Appendix I: Molecular and equivalent weight of some important compounds

Appendix II: Percentage concentration, specific gravity, normality and amount needed for making 1N solution of some commonly used acids

Appendix III: some important conversion factors

Appendix IV: Common terms used in pesticide residue analysis

Appendix V: Classification of pesticides according to toxicity, expressed as LD_{50} (mg/kg)

Appendix VI: Chemical classification of pesticides

Appendix I

Molecular and equivalent weight of some important compounds

S.No.	Compound	Formula	Mol. Wt. (g)	Eq. wt.(g)
1.	Ammonium acetate	$CH_3COOHNH_4$	77.08	77.08
2.	Ammonium chloride	NH_4Cl	53.49	53.49
3.	Ammonium fluoride	NH_4F	37.04	37.04
4.	Ammonium nitrate	NH_4NO_3	80.04	80.04
5.	Barium acetate	$(CH_3COO)_2Ba$	255.43	127.72
6.	Barium chloride	$BaCl_2.2H_2O$	244.28	122.14
7.	Boric acid	H_3BO_3	61.83	20.61
8.	Calcium acetate	$(CH_3COO)_2Ca$	158.00	79.00
9.	Calcium carbonate	$CaCO_3$	100.09	50.05
10.	Calcium chloride (dihydrate)	$CaCl_2.2H_2O$	147.02	73.51
11.	Calcium hydroxide	$Ca(OH)_2$	74.00	37.00
12.	Calcium nitrate	$Ca(NO_3)_2$	164.00	82.00
13.	Calcium sulphate	$CaSO_4.2H_2O$	172.27	86.08
14.	Ferrous ammonium sulphate	$(NH_4)_2SO_4FeSO_4.6H_2O$	392.13	392.13
15.	Ferrous sulphate	$FeSO_4.7H_2O$	278.01	139.00
16.	Magnesium chloride	$MgCl_2.6H_2O$	203.30	101.65
17.	Magnesium nitrate	$Mg(NO_3)2.6H_2O$	265.41	128.20
18.	Potassium chloride	KCl	74.55	74.55
19.	Potassium dichromate	$K_2Cr_2O_7$	294.19	49.04
20.	Potassium hydroxide	KOH	56.10	56.10
21.	Potassium permanganate	$KMnO_4$	158.03	31.60
22.	Potassium nitrate	KNO_3	101.10	101.10
23.	Potassium sulphate	K_2SO_4	174.27	87.13
24.	Potassium hydrogen phthalate	$COOHC_6H_4COOK$	204.22	204.22
25.	Oxalic acid	$C_2H_2O_4.2H_2O$	126.00	63.00
26.	Silver nitrate	$AgNO_3$	169.87	169.87
27.	Sodium acetate (anhydrous)	CH_3COONa	82.04	82.04
28.	Sodium bicarbonate	$NaHCO_3$	84.01	84.01
29.	Sodium carbonate	Na_2CO_3	106.00	53.00
30.	Sodium chloride	$NaCl$	58.45	58.45
31.	Sodium hydroxide	$NaOH$	40.00	40.00
32.	Sodium nitrate	$NaNO_3$	84.99	84.99
33.	Sodium oxalate	$Na_2C_2O_4$	134.00	67.00
34.	Sodium sulphate	Na_2SO_4	142.04	71.02
35.	Sodium thiosulphate	$Na_2S_2O_3.5H_2O$	248.18	248.18

Appendix II

Percentage concentration, specific gravity, normality and amount needed for making 1N solution of some commonly used acids

S.No.	Reagent	Conc. (%, w/w)	Sp. Gravity ($g\,mL^{-1}$) at 20°C	Normality (approx.)	Quantity (mL) required for 1L of *1N* solution
1.	Acetic acid glacial	99.0	1.06	17.5	58
2.	Ammonium hydroxide	28.3	0.91	15.0	67
3.	Hydrochloric acid	39.0	1.19	11.8	78
4.	Nitric acid	71.0	1.42	15.6	62
5.	Phosphoric acid	85.0	1.70	44.0	23
6.	Sulphuric acid	96.0	1.84	36.0	28

Appendix III

Some important conversion factors

ppm	mg/litre or µg/ ml
1 ppm	2.25 kg/ha
10,000 ppm	1 per cent
1 ha	2.471 acres
1 acre	0.405 hectare
1 kg/ha	1.12 lbs/acres
Mesh number	16/ mm
1°A	10^{-8} cm or 10^{-10}m
Organic matter	OC(%)×1.724
P	P_2O_5×0.44
P_2O_5	P×2.29
K	K_2O×0.83
K_2O	K×1.20
SO_4	S×3
CaO	Ca×1.40
Protein (%)	N(%)×6.25

Appendix IV

Common terms used in pesticide residue analysis

ADI	Acceptable daily intake of a pesticide is the daily intake which during the entire life time appears to be without appreciable risk to the health of the consumer on the basis of all the facts known at the time of evaluation of the pesticide. It is expressed in milligram of pesticide per kilogram of body weight.
Chromatography	Chromatography is a laboratory technique for the separation of a mixture.
Clean-Up	After extraction removing of other impurities from the matrix.
Extraction	Extraction in chemistry is a separation process consisting in the separation of a substance from a matrix.
Half life	Half-life is the time required for a quantity to reduce to half its initial value.
LD_{50}	The median lethal dose of a toxin, radiation, or pathogen is the dose required to kill half the members of a tested population after a specified test duration. LD_{50} figures are frequently used as a general indicator of a substance's acute toxicity. A lower LD_{50} is indicative of increased toxicity
Matrix	The component of interest in the sampleis called the analyte, and the remainder of the sample is the matrix.
Metabolite	A metabolite is a substance or compound formed during metabolism, which is an overall set of chemical reactions that occur in an organism or cell.
MRL	Maximum residue limit is the maximum concentration for a pesticide residue on crop or food commodity resulting from the use of pesticides according to good agricultural practices. Its concentration is expressed in mg/kg.
Residue	A small amount of something that remains after the main part has gone or been taken or used.
Solvent	Solvent is a substance that dissolves a solute resulting in a solution. A solvent is usually a liquid but can also be a solid, a gas, or a supercritical fluid.
Toxicity	Toxicity is the degree to which a chemical substance or a particular mixture of substances can damage an organism.
Waiting period	Waiting period is the determination of the minimum interval required between application and harvest of the crop.

Appendix V

Classification of pesticides according to toxicity, expressed as LD_{50} (mg/kg)

WHO class	Classification	LD_{50} for the rat ($mgkg^{-1}$ b.w.)		Label
		Oral	Dermal	
Ia	Extremely hazardous	<5	<50	POISON
Ib	Highly hazardous	5-50	50-200	POISON
II	Moderately hazardous	50-2000	200-2000	DANGER
III	Slightly hazardous	2000-5000	2000-5000	CAUTION
IV	Unlike to present actual hazard	>5000	>5000	-

Appendix VI

Chemical classification of pesticides

Organophosphate	These chemical substances which are produced due to reaction between phosphoric acid and alcohols. *Mode of action*: It affects the nervous system by inhibiting the action of enzyme acetyl cholinesterase (AChE). This causes irreversible blockage leading to accumulation of the enzyme which results in overstimulation of muscles. *Example:*These mainly include insecticides, nerve gases, herbicides, etc.
Carbamate	These are esters of carbamic acids. *Mode of action*: Inhibiting acetyl cholinesterase similar to that of the organophosphates but the bond formed for inhibition is less durable and thus reversible. *Example*: These also include mainly of insecticides.
Organochlorine	These are the derived from chlorinated hydrocarbons. *Mode of action*: These are endocrine disrupting agents which effect on the hormonal systems of the body, act as duplicates of the normal hormones and thus causing adverse health problems. They remain in environment for a long time by breaking down slowly and accumulating in the fat tissues of animals. *Example:*A well-known example is DDT (dichloro diphenyl trichloroethane).
Pyrethroid	These are derivatives of ketoalcoholic esters of chrysanthemic and pyrethroic acids and are more stable in sunlight than pyrethrins. These are most popular insecticides as they can easily pass through the exoskeleton of the insect. *Mode of action*: These are potent nuero poisons, endocrine disruptors and cause paralysis. Pyrethroids are synthetic version of pyrethrin a natural insecticide. They have similar chemical structure and similar mode of action as of pyrethrin which is obtained from chrysanthemum. *Example:*Few examples are- deltamethrin, cypermethrin, etc.

References

Berger, K.C. and Truog, E. (1939). Boron determination in soils and plants. Ind. Eng. Chem. Anal. Ed. 11:540-545.

Black, C.A. (1965). Methods of Soil Analysis, Part I, Physical and Mineralogical Properties of Soil, (including statistics of measurement and sampling) No. 9 in the series Agronomy, *Am. Soc. Agronomy*.

Bouyoucos, G.J. (1927). The hydrometer as a new method for the mechanical analysis of soils. *Soil Sci.*, **23**: 343-353.

Chesnin, L. and Yien, C.H. (1950). Turbidimetric determination of available sulphates. *Proc. Soil Sci. Am.*, **14**:149-51.

Handa, S. K., Agnihotri, N. P., and Kulshrestha, G. (1999). *Pesticide residues: significance, management and analysis*. red leaf fly.

Jackson, M.L. (1973). Soil Chemical Analysis, Prentice Hall of India Pvt Ltd., New Delhi.

Jackson, M.L. (Ed.) (1958). Boron determination for soil and plant tissue. In Soil Chemical Analysis p370-387.

Kulshrestha, G. Dixit, A.K., Gajbhiye, V.T., Mukherjee, I. and Gupta, S. *Pesticide Referral Laboratory*, Training, Division of Agricultural Chemicals, IARI, New Delhi.

Lindsay, W.L. and Norvell, W.A. (1978). Development of DTPA soil test for zinc, iron, manganese and copper. *Soil Sci.Soc. Am. J.*, **42:** 421-48.

Olsen, S.R., Cole, C.V., Watanabe, F.S. and Dean, L.A. (1954). Estimation of available phosphorus in soils by extraction with sodium bicarbonate. U.S. Department of Agriculture Circular, 939, pp 19.

Pesticide Analytical Manual, *volume 1*: Multi residue methods: FDA, 2001.

Pesticide residue in foods by acetonitrile extraction and partitioning with magnesium sulphate, *AOAC official method.* 2007.01

Piper, C.S. (1951). Soil and plant analysis. The University of Adelaide Press, Austalia.

Sharma, K. K. (2007). *Pesticide residue analysis manual*. ICAR.

Singh susheel, Panchal PR, Joshi MN, Litoriy NS and Shah PG. (2012). Development and validation of fast multi residue method for organochlorine pesticides from high fat milk with QuEChERS approach. *Pesticide Research Journal* 24(2):205-211

Stahl E. (1969). *Thin layer chromatography- A laboratory handbook*, Springer-Verlag, Berlin.

Subbiah, B.V. and Asija, G.L. (1956). A rapid procedure for determination of available nitrogen in soils. *Curr. Sci.*, **25**: 259-260.

Walkley, A. and Black, I.A. (1934). An examination of the digestion method for determining soil organic method, and a proposed modification of the chromic acid titration method. *Soil Sci.*, **34**: 29-38.

Zeitfracht Medien GmbH
Ferdinand-Jühlke-Straße 7
99095 Erfurt, Deutschland
produktsicherheit@kolibri360.de